Herbert
Wendt

Book Club Associates

From Ape to Adam

The search for the ancestry of man

291 illustrations, 19 in colour

The drawing by Rudy Zallinger reproduced on the title page
is taken from *Early Man* by permission of Time-Life Books

The extract from *Meet your Ancestors* by Roy Chapman Andrews
on pages 183–4 is reprinted by permission of the Hutchinson
Publishing Group Ltd and The Viking Press Inc. That from *Finding
the Missing Link*, by Robert Broom on pages 217–18, by permission
of Sir Isaac Pitman and Sons Ltd

Translated from the German *Der Affe Steht Auf*
by Susan Cupitt
Selection and layout of illustrations by Thames and Hudson
This edition published 1974 by Book Club Associates
by arrangement with Thames and Hudson Ltd
© Thames and Hudson Ltd, London 1971 and 1972
Original text © Rowohlt Verlag GmbH,
Reinbek bei Hamburg, W. Germany 1971
Typesetting, printing and binding by
Jarrold and Sons Ltd, Norwich
Printed in Great Britain

Contents

Introduction   7

Part One

The question of all questions   13

The ape on the silver bowl 15 / Anatomical studies in Renaissance times 19 / Man is made of the same materials as animals 22 / Were Adam and Eve giants? 23 / The scandal of the faked stones 25 / The Deluge, sinners and salamanders 27 / Linnaeus' System of Nature 30 / Is man a degenerated ape? 33 / Goethe discovers the intermaxillary bone 36 / Buffon explodes the biblical chronology 38 / Lamarck and the inheritance of acquired characteristics 40 / Commemorative medals of Creation 42 / Stone Age, Bronze Age, Iron Age 43 / Fossil man does indeed exist 46 / A thirty years' struggle to prove Adam's ancestors 50 / 'The Chancellor of the Exchequer among Scientists' is won over 53 / Controversial bones from the Neander valley 55 / At last gleams of light 58 / A world trip makes history 60 / The struggle for existence 63 / Is man an ape or an angel? 65 / Darwin's bulldog 69 / Sexual selection 71 / The ape-complex 73 / The history of Creation 77 / The speechless ape-man 82

Part Two

Artists of the Ice Age   85

Groans about the Ice Age 87 / The paradise of prehistoric man 89 / A case of prehistoric murder 90 / Hand-axe cultures in Europe 94 / A dog disappears down a crevice 96 / Maria and the bulls 97 / Henri Breuil breaks the ban 103 / Are there forged cave paintings? 107 / Cult and magic in the Ice Age 110 / Four youths find a prehistoric Louvre 114 / Sex 20,000 years ago 118 / Late Ice Age expressionism 122 / The Grimaldi caves 126 / Eating someone out of love for them 130 / A Pompeii of prehistoric times 132 / Justice for the Neandertalers 134 / Prehistoric men at places of Biblical interest 138 / A 'modern' European lived 300,000 years ago 140.

Part Three

The dawn of humanity 147

The Piltdown skull 149 / The riddle of Broken Hill 155 / The clocks of the earth 156 / The Mauer mandible 160 / The oldest hearths of humanity 163 / Head-hunters on the Solo River 173 / Searching for early men in Java 176 / A Jesuit starts a ball rolling 180 / The case of the dragons' teeth 182 / Here lies primitive man 183 / Celebrating the discovery of Peking man 186 / Where is Nelly? 192 / There were once giants on earth 193 / The trail takes us to Africa 198.

Part Four

Between animal and man 203

The Missing Link sensation 205 / 'Dart's child' criticized 209 / Broom and the African story of Creation 212 / When a boy collects bones 216 / The Missing Link no longer missing 219 / 'Mrs Ples' and the killer apes 221 / Like a report from a morgue 224 / More than a hundred australopithecines 228 / Olduvai – a unique site for fossil finds 230 / Right back in the Tertiary era 236 / The highest seat of anthropological judgment 238 / Australopithecines as toolmakers 240 / The dexterous pygmy 245 / Our embarrassing relations 249 / The Naked Ape 251 / Angel of the chimpanzees 252 / A genuine 'Aha' experience 254 / Dehumanized apes 256 / The mountain ape of Monte Bamboli 260 / Twenty million years of past time 266 / Where does man begin? 267 / The problem of brain development 270 / Where the ways of Cain and Prometheus part 274.

Epilogue   276

Bibliography   278

List of Illustrations   280

Index   285

Introduction

One of the greatest, most fascinating, and perhaps most disillusioning discoveries man has ever made is the discovery of his own origin and racial history. This study of our prehistory is still young as a subject of research. It is one of the scientific 'break-throughs' of the nineteenth century, a child of modern humanism, born when our study of the natural sciences began; during the last hundred years man has gradually, albeit somewhat reluctantly, come to accept that he too is part of nature.

Modern anthropology has meant the end of Adam and Eve; it has opened up new worlds to us, forced upon us a radical rethinking such as man has not had to face since the days of Copernicus and Galileo. It has brought about an earth-shaking revolution, the effects of which can still be felt today. But this conflict of views has been a fruitful one; for despite its moments no less of farce than of drama it has greatly advanced the history of our Western culture, and enriched it too. Today our concern about the problem of our origins seems perhaps more serious and more urgent than it did to those who lived a hundred years ago, for behind the question, Where do we come from? there looms another, with its undertones of menace, Where are we going? And the one is unanswerable without the other. Our knowledge of man's present-day existence would lack all sense if it were unrelated to our prehistoric past, or if we had no knowledge of it.

TAB. XXIII.

GENESIS cap. I. v. 26. 27.

Homo ex Humo.

I. Buch Mosis cap. I. v. 26. 27.

Erschaffung und Zeugung deß Menschen.

This pictorial history of the early development of man began to take shape as I prepared my earlier book, *I Looked for Adam*. It recorded the history of a science: recounting historical events, discoveries, excavations and scholars' achievements took precedence over the presentation of what had actually been found and the documents relating to these finds. Here, on the other hand, I shall focus attention upon the basic elements of modern anthropology; the documentary evidence should for the most part speak for itself.

For prehistoric finds cannot be satisfactorily described in words alone; they must be *seen* before one can begin to understand them. On the other hand if they were merely set out like museum pieces they would be lifeless. Only by portraying them in both words and pictures can we bring them to life again. In this book, therefore, the illustrations and the text, the quest, the inquirer and his subject-matter are intended to form an integral whole, so far as that is possible and artistically permissible. A documentation of this sort in words and pictures provides the reader with the easiest means of transporting himself back into the past, and of understanding how the subject he is studying has developed. The intention is to relate the various prehistoric finds and documents one with the other by the order in which they are presented and by the linking text. Indeed, the many different stages in the development of anthropology are not stories in themselves, but a continuing history, an indivisible whole.

Since *I Looked for Adam* was first published, there have been so many new discoveries and interpretations, particularly on the subject of how man evolved, that the text has to some degree outrun the illustrations. The bold and interesting theories of controversial authors such as Desmond Morris and Robert Ardrey require just as much discussion as the revolutionary discoveries made by Louis Leakey in East Africa, discoveries which make it absolutely clear that we must extend the age of the hominids, that is, those mammals from which we are derived, by some fifteen or twenty million years.

Opposite: 'The Creating and Begetting of Mankind', copper engraving from J.J. Scheuchzer's 'Physica Sacra', 1771

It is obviously impossible to make a really exhaustive study of a theme so far-ranging that it often intrudes upon other areas of natural history and the history of our intellectual development. However much the author tries to let the facts speak for themselves, by the mere act of selecting from the material he has at his disposal he sets his own stamp upon it – the effect of his own approach to the material and, indeed, his personal involvement. This literary compromise is, I believe, quite legitimate; it is the only means we have of unlocking the secret treasuries of science and of providing the reader with a basis for gaining further knowledge of mankind's past on earth.

A book which aims at integrating the scientific theories and achievements of many centuries can only be brought to fruition if the author has gained an extensive knowledge of the materials available. I would therefore like to thank all those people and institutions who have assisted me in obtaining material and have given me the benefit of their valuable advice. In particular I would like to name the following:

Professor Dr K. Absolon, Brno; Museumsdirektor K. Brandt, Herne (Westphalia); †Abbé Henri Breuil; Professor Dr Irenäus Eibl-Eibesfeldt, Seewiesen; Professor Dr Wilhelm Gieseler, Tübingen; Professor Dr Gerhard Heberer, Göttingen; Professor Dr Edwin Hennig, Tübingen; Professor Dr G. H. R. von Koenigswald, Frankfurt am Main; Professor Dr Herbert Kühn, Mainz; Professor Dr Konrad Lorenz, Seewiesen; Dr Kenneth P. Oakley, London; Professor Dr Adolf Portmann, Basle; †Professor Dr Hans Weinert, Heidelberg; Dr Wolfgang Wickler, Seewiesen.

I have derived much profit, furthermore, from reading the works of Roy Chapman Andrews, Robert Ardrey, Josef Augusta, Geoffrey Bibby, †Robert Broom, Sonia Cole, Raymond A. Dart, Adriaan Kortlandt, Jane van Lawick-Goodall, Louis S. B. Leakey, †Wilfred E. Le Gros Clark, Desmond Morris and †Pierre Teilhard de Chardin; to these I am indebted, as well as to the following museums and institutions, libraries, and scientific societies, which have been most helpful: Anthropologische Forschungsstelle,

Göttingen; Badische Landesbibliothek, Karlsruhe; British Museum, London; British Museum (Natural History), London; Emschertal-Museum, Herne (Westphalia); Hugo-Obermaier-Gesellschaft, Erlangen; Institut für Anthropologie und für Vorgeschichte, Tübingen; Max-Planck-Institut für Verhaltensphysiologie, Seewiesen; Musée de l'Homme, Paris; Musée des Antiquités Nationales, Saint-Germain-en-Laye; Musée National de Préhistoire, Les Eyzies; Senckenberg-Bibliothek, Frankfurt am Main.

The support of my publishers, Thames and Hudson Ltd, London, and Rowohlt Verlag, Reinbek, near Hamburg, has far exceeded the assistance that publishers usually render. It is in great measure thanks to their unusually ready and sympathetic cooperation that the plans for this book have been realized at all.

Baden-Baden, August 1970   H. W.

Part One

The question of all questions

The greatest mystery of all is man himself.
Finding a solution to the riddle he presents
is the endless work of world history.
                                        NOVALIS

LEONARDO DA VINCI

SCHEUCHZER

GOETHE

BUFFON

LAMARCK

CUVIER

THOMSEN

BUCKLAND

PENGELLY

BOUCHER DE PERTHES

LYELL

FUHLROTT

DARWIN

T.H. HUXLEY

VOGT

HAECKEL

On the outer frieze (towards the
bottom) of this silver bowl a great
ape is portrayed, erect and armed.
(Detail in drawing below)

A silver bowl from an Etruscan grave is the earliest evidence we have today of an encounter between civilized man
and his nearest relations in the animal world. It was found
in ancient Praeneste, in Latium, and according to the
archaeologist Ludwig Curtius is part of a whole series of
dishes of Phoenician and Carthaginian origin. Its outer
frieze consists of an almost cinematographic series depicting
a most remarkable being, which can be seen most clearly
in the illustration on the right. This creature walks on two
legs, is clothed in hair, is solidly built, and has a massive
skull. It is carrying a stick in one hand and is throwing a
stone with the other. The artist seems to have known this
being well; for both its physical proportions and the typical
attitude of attack leave no room for doubt that it is a great
ape, perhaps a gorilla, which is depicted here.

Similar figures are portrayed in the work of other ancient
cultures, such as the killing of a 'wild man' on a dish from
Cyprus, a Babylonian terracotta of the demon Chumbaba,

with the head of an anthropoid ape, and many representations of man-like ape-gods from ancient India. That the great apes, at least, were considered not animal but half-human in the ancient Orient is evident from a report written about 525 BC by the Carthaginian mariner Hanno, which is the earliest authentic but by no means sole evidence we have of encounters with such creatures. Hanno tells how he came upon 'forest people in animal skins' on the coast south of the present-day Cameroons, 'among them many women with shaggy heads whom our interpreters called gorillas'. The legend of the furred man Eabani, to be found in the Epic of Gilgamesh, has its significance too. The demi-god Gilgamesh kneads Eabani out of a forest animal and goes on to conquer the world with him as his companion.

These ancient pictures and reports betray the first inkling of a belief in a wild, animal-like, primitive form of man, or at least in some relationship between ape and man. Such ideas did not wholly disappear even in Greek and Roman times; but because they did not fit in with the philosophical and artistic ideas of the age they became relegated more and more to the world of mythology. Man is the measure of all things, declared the Sophist Protagoras of Abdera as early as 450 BC. He is the desired aim of creation, according to Aristotle. His natural being is changeable and transitory, added Plato; only the world of ideas, that is, the world of the good and the divine, he maintained, is real and lasting. The growing predominance of anthropomorphism and the Platonic doctrine of ideas meant that the 'animal-men' became increasingly relegated to the mythological realm, where they became giants, monsters, satyrs, cyclops and gorgons.

The ancient legendary being, the Gorgon, is of particular interest for our investigation. The oldest pre-Corinthian versions of the horrific Gorgon mask have dagger-like teeth, a protruding tongue, and sometimes even a beard. Only later was it transformed into a female head girt about with snakes. Writers of antiquity such as Diodorus and Alexander of Myndos tell us that this monster, half animal, half human, was not merely the product of the imagination

but was derived from a real being. Thus the Gorgon was based on a wild, shaggy form of life which came from Africa and had strength enough to strangle leopards. The sort of being with whom the ancient Greeks associated it can be deduced from the fact that they applied the name 'Gorgadas' to the three gorilla skins Hanno brought back with him from Africa and which were displayed in the temple of Melkarth in Carthage.

But the thinkers of antiquity were not content with an imaginary world of fable; they had the courage to speculate

The Gorgon, seen here painted on a Greek vase, was probably originally based on the face of an anthropoid ape

17

Ancient Asia was already aware in antiquity of the link between ape and man, as this statue of the Ceylonese monkey-god Hanuman shows.

on the origin of man as well. Heraclitus considers animal species to be mutable, that is, subject to change; Animaxander sees man as having arisen from the animal world; Empedocles already presents nature as selective, destroying the weak and preserving only the strong, thus foreshadowing Darwinism in a superficial sort of way. Important Greek and Roman scholars such as Megasthenes and Pliny took up the ancient Oriental terminology again and elevated apes to 'wild men'. In the first century BC the Roman poet and anthropologist Lucretius, the author of perhaps the most important didactic poem of antiquity, went a step further: he declared that man's ancestors were not gods but wild, animal-like beings who first fought for their survival with tooth and claw, then later with sticks and stones.

The evolutionary biologists of Darwin's time were not entirely mistaken in recalling classical antiquity's intimations of man's true origin. Though indeed it is difficult to make out many details in the surviving fragments of their writings, it is probable that the Greek, Roman and Oriental thinkers, natural historians, artists and chroniclers were more aware of the animal origin of man than were those of any subsequent period till the time of the Renaissance, indeed perhaps until well into the seventeenth and eighteenth centuries. Their early tentative advances into the hazy realm between animal and man came to an end when Platonism overran the Western intellectual world. In the words of the American biographer of Darwin, Professor William Irvine: 'The great historical enemy of evolution has been the Platonic tendency – so congenial to logic, morals and mathematics – to regard the universe as a fixed order, in which realities remain perspicaciously what they are while the mind thinks about them.'

Nature, then, was no longer thought to be in a continual state of flux, as Heraclitus had taught, but was a system in which everything had its divinely ordained place, classification and purpose – a great 'chain of being' which extended in strict linear gradation from lowest to highest. All types and forms were mere copies of external and immutable ideas; they were as eternal and unchanging as the ideas from which they were derived. An unbridgeable gulf was formed between man, the pinnacle of Creation, who had been made in the image of God, and the rest of the world, separating them completely. Almost two thousand years of stagnation had to pass before the inspired men of Renaissance medical science found the courage to reconsider antiquity's superficial theories of biology.

Khnemu fashions the first man out of clay. Drawing based on an Egyptian relief

'Would that it might please our Creator that I were able to reveal the Nature of Man and his Customs even as I describe his Figure.'

Surprisingly enough this statement, which so echoes the attitudes of modern medicine, is to be found in some notes on anatomy written between 1493 and 1503. Their author, Leonardo da Vinci, was a sort of universal man, a genius whose versatility made him an exception even among the exceptional men of the Renaissance: painter, sculptor, architect, physicist, engineer, palaeontologist, philosopher, writer, anthropologist, and comparative anatomist. And it was comparative anatomy, which brought anthropology to the forefront of interest in Renaissance times.

During the Middle Ages the dissection of human corpses was forbidden, so that any knowledge of the anatomy of man had to be gained from the dissection of apes. Leonardo was one of the first to dissect the human body in secret, and

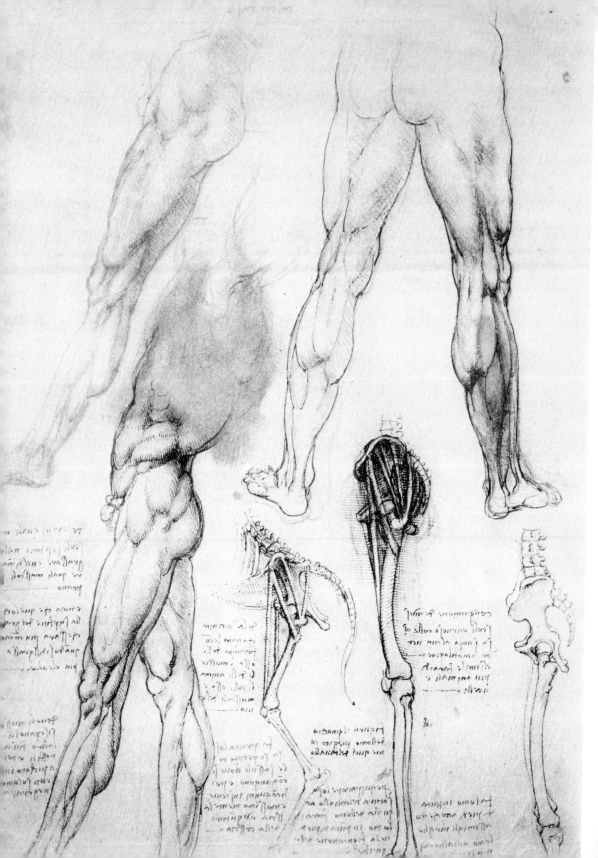

the numerous similarities between man and ape are implicit in his injunction: 'Describe man, including such creatures as are of almost the same species, as Apes, Monkeys and the like, which are many.' The conclusion he drew from his anatomical studies was incredibly bold for his time: 'Man in fact differs from animals only in his specific attributes.'

Shortly after Leonardo's notes on anatomy had been completed, a man whom modern anatomists still honour as their spiritual forefather, was born in Brussels: Andreas Vesalius, later personal physician to Charles V and Philip of Spain. His famous work *De humani corporis fabrica* appeared in 1543. We reproduce the title-page (below left), showing a human

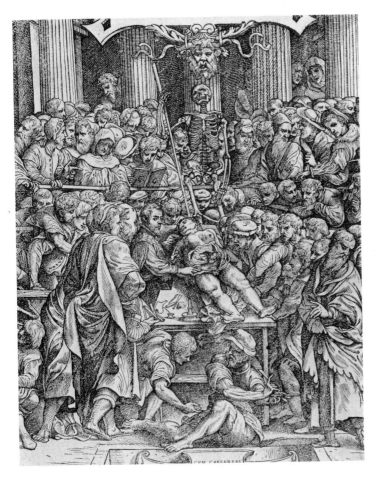

Opposite: Drawings by Leonardo da Vinci. A study of the anatomical difference between the hind leg of a horse and the human leg (position of the knee and heel joints). Left: Vesalius dissecting in the anatomy theatre. Below: wood engraving from 'De humani corporis fabrica' by Andreas Vesalius

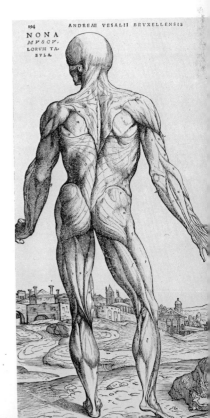

body being dissected by Vesalius in the anatomy theatre of Padua University. The anatomical wood engravings in Vesalius' book are the work of Calcar, pupil of Titian. They eradicated all sorts of established errors and were unsurpassed until quite recent times.

Anatomically, then, man should be classed as an animal. This ancient piece of knowledge, accepted by the natural philosophers of Greek and Roman antiquity, was finally reaffirmed by the anatomists of the Renaissance.

'Man is fram'd of Materials not exceeding in value those of other Animals', wrote the French philosopher Julien Offroy de la Mettrie, two hundred and fifty years after Leonardo. He saw the moral implications of the knowledge gained from recent medical research: 'There is, they say, a law of nature, a knowledge of right and wrong deeply imprinted on the mind of man, which in other animals is not perceived. But this law is natural to animals as well. We know that we think and that we are stricken by remorse after we have done any guilty action, and certain indications lead us to conclude that other men do so too. But these same indications are also to be observed in animals.'

La Mettrie, one of the first exponents of philosophical materialism, did not consider even speech a criterion of human reason. The art of speech, he declared, was learned through practice and upbringing. He even quite seriously considered teaching an ape to speak by the methods used to teach deaf-mutes. But the thoughts and theories of this early pioneer of evolution were not to be taken up. It was only later, when fossils became the focus of interest and when men reflected upon these remains of prehistoric beings that they began to think again about the origin of the human race.

Julien Offroy de la Mettrie, founder of Materialism, tried to teach an ape to speak

While Swedish troops were digging siege trenches at Hundstieg near Krems in Lower Austria in 1645 they came upon a number of large, worn and partly broken bones. Two years later the distinguished topographer, Matthäus Merian the Older, described this find. The soldiers, so he said, had 'found a giant body of tremendous size, in a patch of ground made dark by the decay of flesh. The head and bones were much damaged and dispersed, for they were quite worn by age and decay, mouldy and very brittle. So many bones were found that learned and experienced men who saw them judged them to be human bones.'

Indeed, fossilized teeth and bones had been constantly turning up since antiquity. Sometimes of surprising size, they were generally considered to be the remains of giants or similar monsters, just as Merian had supposed. In most cases they belonged to mammoths and other great creatures of the Ice Age, and many of these 'giant bones' were displayed as curiosities in castles, palaces, churches, monasteries and town halls. In 1577 a collection of mammoth bones even evoked a violent theological argument. It was thought at first that these relics should be given a Christian burial and laid in the city cemetery, but Dr Felix Platter of Basle studied them closely and declared them to be the bones of a giant eighteen feet tall, who as a heathen merited no such consideration.

In 1678 the Jesuit father and versatile scholar Athanasius Kircher made an intensive study of all then known fossils. Kircher was the first speleologist, or explorer of caves; but

In the seventeenth century these mammoth teeth, depicted by M. Merian in his 'Theatrum Europaeum', were thought to be the teeth of giants

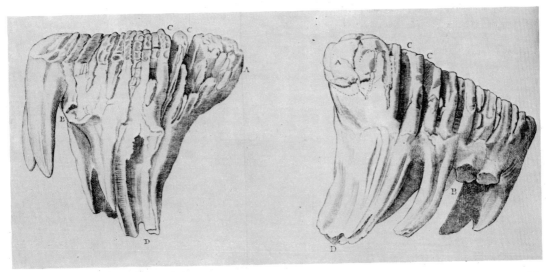

as a child of his time he took the existence of giants and monsters for granted. In his book *Mundus subterraneus* Kircher describes four different sorts of 'giants'. Other scholars even solemnly debated whether Adam and Eve had not perhaps been giants. Henrion, a Frenchman of the seventeenth century, calculated on the basis of existing 'giant bones' that Adam had been 123 feet 9 inches and Eve 118 feet 9 inches tall.

Thus the theory of fossilized 'giants' led the study of prehistory astray. Nevertheless, the mere discovery of these bones aroused a general interest in the fossilized remains of a past world, and the question, Were there people in those ancient times? came to preoccupy men's minds. This question has haunted palaeontologists ever since.

Three hundred years ago it was thought man was descended from giants. Common man (Homo ordinarius) is shown here alongside four 'giants'. Taken from 'Mundus subterraneus' by Athanasius Kircher, 1678. Right: 'Giant bone', found 1443, and engraved with the motto of Emperor Frederick III (AEIOU), and the year of its discovery; the portal of the Cathedral of St Stephen in Vienna, which was the repository of this thigh bone of a mammoth over a long period, came to be known as the 'Giant's Portal'

Gigantis Sceleton
in monte Erice prope Drepanum inventum Boccatio *teste 200 cubitorum.*

A zealous collector of fossils was Johannes Bartholo-
mäus Adam Beringer, university professor and councillor of
the city of Würzburg. A product of the early eighteenth
century, Beringer felt himself hemmed in by the restric-
tions imposed by medieval thinking, which regarded fossils
as mere freaks of nature, the products of a mysterious *'vis
plastica'* – a playful force which had the power to fashion all
manner of animal and plant shapes in stone. What happened
to Beringer exposed the full absurdity of this doctrine.

Between 1724 and 1726 Beringer dug up a great number
of strange and often very realistic-looking 'fossils' from the
shell-lime near Eibelstadt: flowers and fruit, snails, crabs and
scorpions, insects and fish, spiders in their webs, coupling
frogs, lizards, salamanders and birds, also imprints of suns
and moons, stars and comets, indeed even Hebraic inscrip-
tions bearing the word Jehovah. Beringer was entirely convinced of the authenticity of his finds, and as he had found
about two thousand of these 'picture stones' he considered
they must have some scientific value. So, with his pupil
Georg Ludwig Huebes, he published the results of his
investigations in a large and splendid volume, profusely
illustrated, entitled *Lithographiae Wirceburgensis*.

But the colleagues whose support Beringer enlisted were
not so easily convinced. They perceived the marks of chisel
strokes upon the stones, and suggested that the three
Eibelstadt youths who had encouraged Beringer to start
collecting had simply faked them and buried them for
Beringer to find. Beringer was deeply incensed at such a suggestion, and immediately sought the opinion of a Jesuit to
confirm their authenticity.

Later it turned out that the distinguished historian and
prehistorian Johann Georg von Eckardt, who felt some
personal animosity for Beringer, had played a malicious
trick upon the latter. It was he, together with his assistant
Ignaz Roderique, who had persuaded the three boys to carve
and bury the stones; indeed Roderique may well have been
responsible for many of them himself. But even though this
trick was discovered while Beringer was still alive, many
collectors of curiosities and fossils remained firmly convinced

25

up to some forty years later that these Würzburg stones were genuine. Beringer himself, to the day of his death, could not bring himself to accept that his stones were mere crude forgeries; yet he is reputed to have tried to get the whole edition of his book destroyed.

Soon after this scandal another doctrine came into vogue which was warmly greeted by the Churches because it so admirably fitted the story in *Genesis*: the theory of a Universal Deluge. Its most ardent and distinguished advocate was the Zurich city doctor and canon Johann Jakob Scheuchzer.

'The bony skeleton of one of those infamous men whose sins brought upon the world the dire misfortune of the Deluge', was Scheuchzer's description of a fossil he discovered in 1726 near the small village of Oeningen on Lake Constance. Today we know the bones to be those of a giant extinct salamander of the Tertiary. But, influenced by the writings of an Englishman, Professor John Woodward, Scheuchzer believed all fossils to be the result of a catastrophic universal deluge, which was identified with the biblical Flood. This theory caused him to wonder how things had been before the great catastrophe took place, and to search for 'antediluvial' men as well.

Opposite: The fake fossil stones of Würzburg

Scheuchzer's 'afflicted skeleton of an ancient sinner' – actually a Tertiary giant salamander

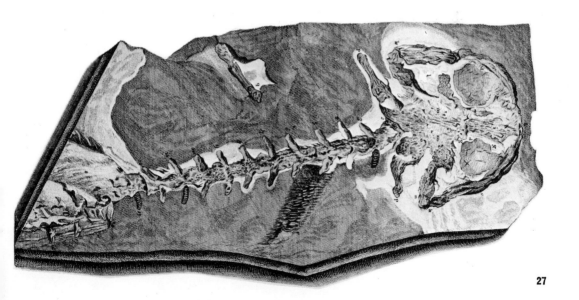

Johann Jakob Scheuchzer, from his 'Physica Sacra'

Scheuchzer described his giant salamander discovered in Oeningen to the English naturalist, Sir Hans Sloane, whose collection was later to form the basis of the British Museum: 'You will like to know, my learned friend, that we have obtained some relics of the race of man drowned in the Flood. Hitherto all that my quite extensive collection had contained were two polished black fossilised dorsal vertebrae. But now my museum has been enriched by the happy discovery of some remains found in the stone of Oeningen, remains which seem to require our most earnest attention. What we have here is no vision of the mere imagination, but the well-preserved bones, and much in number, of a human skull, quite clearly distinguishable from the bones of other species.'

Using the details given in the Bible, Scheuchzer went so far as to date his 'diluvial man'. *Homo diluvi testis*, as he called him, had died in 2306 BC precisely. And to make quite clear that the Flood was to be considered the judgment of God, Scheuchzer's friend, Deacon Miller of Leipheim, composed a couplet about the Oeningen find, which was soon to enter the annals of unconscious humour:

> Unhappy evidence of past transgression,
> Let these stones move the wicked to contrition.

Scheuchzer's pupils and followers called themselves Diluvians, *diluvium* being the Latin word for flood. By stimulating the study of discovered fossils they indirectly fostered the study of prehistoric times, for they saw to it that fossils were systematically collected and classified. Their untiring activity meant that fossils became recognized for what they are, not evidence of the capriciousness of nature, but the petrified remains of creatures who had at some time really walked the earth.

In his 'Museum Diluvianum' Scheuchzer tried to provide scientific evidence of the Deluge legend

MUSEUM
DILUVIANUM

Quod poſſidet

JOH. JACOBUS SCHEUCHZER.
Med. D. Math. Prof. Acad.
Leopoldino-Carolin. Socc.
Regg. Anglic. & Boruſſ.

TIGURI,

Typis HENRICI BODMERL
M D CC XVI.

Sumtibus Poſſeſſoris.

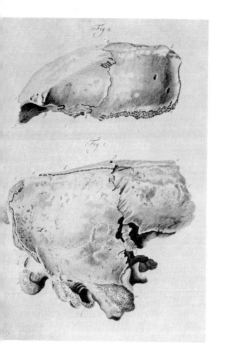

When Johann Friedrich Esper found these fossil bones (probably of cave bears) he concluded that man must have been coeval with 'antediluvial' beasts

'With awed delight', Pastor Johann Friedrich Esper, after finding various skull fragments of Ice Age creatures, in 1771 finally dug up a lower jaw and shoulder blade of apparently human origin in a cave near Gailenreuth near Bamberg. Esper was also a Diluvian, and considered these bones, both animal and human, to be the remains of creatures who had perished in the Flood. Nevertheless, he was the first man to use the knowledge he gained from his exploration of caves to arrive at some conclusion about the life of prehistoric man. If the bones of men were to be found alongside those of animals in the clay of the cave, then 'man must have lived at the same time as these animals'.

Surprisingly enough Esper's book went quite unnoticed. We do not even know whether this Frankish cleric did in fact actually discover the remains of a late Ice Age Cro-Magnon or Neandertal man, or whether he had merely come across some recent human bones in the cave. The first really radical advances in scientific knowledge were made by the great naturalists of the eighteenth century. The earlier rather naïve search for antediluvial men seems to have petered out of its own accord.

Apes and men are the 'master animals'. They are classified together in the mammalian order of Primates – the highest-ranking order of creatures.

It was not a scientific revolutionary who postulated this, but a man whom historians generally consider to have been conservative in his opinions. The Swede, Carl Linné, or Carolus Linnaeus, the first great classifier of nature, maintained that forms were immutable and that his *Systema Naturae*, his classification of the natural world, applied to all periods of time. Thus there was not much he could make of fossils. He dismissed the numerous finds of fossilized creatures which had been made by his time in one mere page of his lengthy work. Scheuchzer's 'ancient sinner', together with the bones of Ice Age creatures, fossilized mussels, snails and others, are relegated to the realm of stone without so much as a comment; he makes no mention of the fact that

Carl von Linné, or Linnaeus, (here portrayed in Lappish national costume) classified the natural world and called man a 'master animal'

the scientific world had for some time regarded fossils as being of organic origin.

But in apparent contradiction to what he thought about fossils, he was firmly convinced of the close relationship between man and ape. His descriptions of this relationship reveal the acuteness of his zoological perception, and his caution too: 'I am well aware of the vast difference between man and animal, when considered from a moral viewpoint.

Man is the only creature whom God has blessed with an immortal soul. But as a naturalist I am concerned with other aspects of his function, and in my study of these I find it most difficult to discover one attribute by which man can be distinguished from apes, except perhaps in the matter of his canines. . . .'

But Linnaeus did not leave it at that: although very little scientific knowledge existed about them, he endowed the anthropoid apes of his time with the Latin names of *Homo troglodytes*, *Homo satyrus*, and *Homo nocturnus*, thus classifying them with man. But there was nothing revolutionary about such an idea, for nearly all the scientists of his time concerned with the subject, whether generally or medically, were of the same opinion. Right into the second half of the eighteenth century we find that works of natural history nearly always depict anthropoid apes as hairy, two-legged creatures with quaintly human expressions upon their faces and a remarkably human body.

The Anthropomorpha, apes with human form, as Hoppius, a pupil of Linnaeus, saw them (1760)

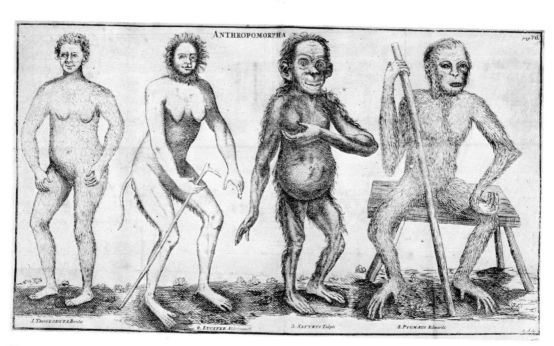

The anatomist Nicolaus Tulp, the central figure of Rembrandt's famous painting, called chimpanzees 'forest men'

Anatomy of the 'Pygmie', from Edward Tyson's 'Orang Outang, sive Homo Sylvestris' (1699). It is in fact, that of a chimpanzee

'They generally walk upright and behave much like other people, but they can also go on all fours.' This was how the Dutchman Dr Jacob Bontius had described orangutans long before Linnaeus' time. His fellow-countryman and eminent anatomist, Nicolaus Tulp, dissected a dead chimpanzee in 1641 and described it as a 'forest man or Indian satyr'. Twenty years later the English anatomist Edward Tyson wrote the first thorough treatise on the anatomy of the chimpanzee, classifying the animal as a 'pygmie'. According to Tyson: 'Our Pygmie does so much resemble a Man in many of its Parts, more than any of the Ape-kind, or any other Animal in the World that I know of.'

Thus, at the end of the seventeenth century, the idea of the missing link was born: an acknowledgment that a link between man and animal did indeed exist, although nothing much was yet known about it. And the animals which it was thought might supply the answer were the same ones the ancients had considered – the forest 'men' or 'wild men', that is, the great apes.

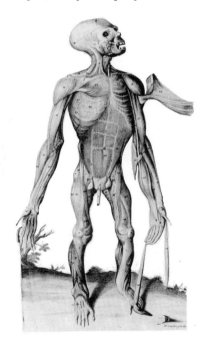

The picture of 'rough-haired Esau' in Scheuchzer's four-volume *Physica Sacra* reveals that 'forest men' even found their way into the works of the Diluvians. Here the ape of Nicolaus Tulp's *Observationes medicae* is represented beside the biblical Esau.

But it was not long before the apes were again rejected as the possible link between men and beasts. Linnaeus' great rival, the Comte de Buffon, had made a thorough study of a young chimpanzee living in the Jardin du Roi in Paris, and although he supported the idea of evolution generally and considered it quite possible that there was some link between ape and man, he could not believe that the animal in his care was in any respect human, and banished it categorically to the realm of apes. But in confidential talks with his closest friends Buffon indicated that in his opinion Linnaeus had not got it quite right: apes were not some kind of backward men, but, on the contrary, man was a 'degenerated ape'.

In Scheuchzer's 'Physica Sacra' one of Nicolaus Tulp's 'forest men' sits beside the 'rough-haired Esau'

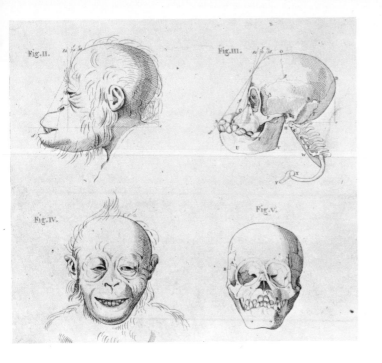

'Anatomical studies of the Orang-utan', by Petrus Camper

'Only man has his incisors in an extended maxilla without any intermediary bone.' This, in the view of Petrus Camper, a Dutch artist and anatomist was the essential difference between man and ape. Dissecting orang-utans, he had noticed that the ape he was studying, like all mammals, had a special bone, the *os intermaxillare*, or intermaxillary bone, between the two halves of the upper jaw. But this bone, as every expert knew, was no longer distinguishable in man.

Unfortunately Camper's study was limited to orang-utans, and he knew nothing about chimpanzees, with the result that he thought the presence or absence of this bone was the most obvious criterion for distinguishing between man and beast; but although this bone is clearly evident in the orang-utan it is as difficult to find in chimpanzees as it is in man. Thus the modest *os intermaxillare* was suddenly promoted into one of the chief means of proving that man was not an animal.

'Something marvellous has happened to me: I have made an anatomical discovery – a fine and important one. You must share my pleasure. But tell no one of it. I've written to Herder about it as well, in the strictest confidence. I'm so happy that I tremble with delight.'

The anatomical discovery which Goethe, whose manifold interests included natural science, wrote about so enthusiastically in his letter of March 1784 to Frau von Stein was the intermaxillary bone. And his letter to Herder ran: 'What I have found is neither gold nor silver, but something which makes me unspeakably happy – the *os intermaxillare* in man! I compared the skulls of men and animals, found a trace, and see, there it is . . . it is like the keystone to mankind, not missing, really there. . . .'

Goethe's famous essay about the intermaxillary bone, often quoted by later evolutionists as a landmark on the way to discovering man's origin, begins with the words: 'The

Goethe, who discovered the inter-maxillary bone in man, thus acknowledging the close link between him and anthropoid apes

close relationship of the ape to man has forced naturalists to the most painful deliberation, and the admirable Camper believed he had discovered that the difference between ape and man was that the former possessed an intermaxillary bone whereas the latter did not.' As Goethe believed that nature was a whole, an organic unity, he did not believe that Camper could be right and was convinced that the *os intermaxillare* could be found in man as well. He made some very thorough studies of bones, and indeed actually found traces of an intermaxillary bone in man, albeit only recognizable as a suture.

The experts scorned 'Goethe's bone'. It was only many years afterwards that the leading German anthropologist Johann Friedrich Blumenbach of Göttingen admitted that Goethe had been right. Blumenbach even toyed with the idea that man was perhaps much older than the biblical chronology led one to assume. He once wrote to Camper: 'There is no conceivable reason why fossil human bones should not be found in the top levels of our globe, just like the fossil bones of elephants and rhinoceroses.'

'This admirable man has a clear and candid insight, a great love of life and delight in all living things: he is interested in all that comes his way.'

Thus Goethe began his literary portrait of the Parisian who created the basis for a new biological theory of the development of mankind. This was Linnaeus' rival, the Comte de Buffon, who worked for over half a century until 1788 in the Jardin du Roi in Paris, later renamed the Jardin des Plantes. It was here, at this institute, that the theory of evolution achieved its first success, but where too it suffered that first bitter defeat from which it was not to recover for a further fifty years.

Buffon was a legend even in his own lifetime. Rousseau, visiting this *grand seigneur* of natural history, was so affected by the experience that he kissed the threshold of the great man's room. Mirabeau considered him the most important man of the century. Voltaire called him a second Archimedes. His gifted pupils, Lamarck, Cuvier and Geoffroy Saint-Hilaire, worshipped him as a god. And Buffon himself was hardly a modest man. Once he declared there were only five great men in the history of mankind: Newton, Bacon, Leibniz, Montesquieu – and himself.

Buffon was the first natural scientist to reject the biblical chronology of early history. The earth had not been created in 4004 BC, as Archbishop Ussher of Armagh and other medieval scholars had reckoned from Old Testament information, but was at least 75,000 years old. And just before his death he came to believe that even this figure was a far too modest one. He drew up a brief history of the development of the world, declaring that: 'The structure of all great works of nature shows us that they can only have arisen by

Paris was the birthplace of modern natural science. Here the reproductive organs of a bitch are being studied under the microscope. Buffon is shown on the extreme right

Buffon (centre) at a gathering of natural scientists in the Jardin du Roi

a slow succession of regular and continual movements.' Creatures adapt themselves to these changes; they develop in accordance with their surroundings and pass on the new characteristics they acquire to their offspring by a process of hereditary memory.

A well-informed pupil of Buffon's, who had led a varied life as a pupil to the Jesuits, discharged army officer, humble clerk, amateur botanist and literary hack, adopted his master's theories and carried them to their logical conclusion. Though descended from the once distinguished line of the Barons of Saint-Martin du Picardie he was impoverished, and had come down much in the world. His name, long and sonorous, was to make world history: Jean-Baptiste Pierre Antoine de Monet, Chevalier de Lamarck.

The first theory of evolution to have any substance was advanced by Jean Baptiste Lamarck

'Species cannot be distinguished completely from each other; they pass into one another, proceeding from simple Infusoria right up to man.' This was Lamarck's first sketchy definition of a new theory of evolution, which he propounded in 1802. Seven years later this radical, who as a result of the events of 1789 became the director of the Jardin des Plantes in Paris, that venerable institute where Buffon had lorded it for so long, stated the case more succinctly:

'Every observant and cultivated person knows that nothing on the surface of this earth remains forever the same. Everything undergoes in time the most gradual changes which take place at varying degrees of rapidity, depending on its own nature and circumstances. . . . These changing environmental conditions bring about a change in the requirements, customs and manner of living of animals, which in turn results in a transformation and development of organisms. Thus these are. subject to imperceptible change, even though such change only becomes noticeable after a considerable period of time.'

Thus according to Lamarck the permanent use or disuse of an organ, determined by the creature's environment, ultimately leads to the transformation of that creature. Animals and plants are the product of their milieu; and the

characteristics they acquire in the course of their adaptation to their surroundings are inherited by further generations, which thus undergo a gradual process of advancement. And the development of man, according to Lamarck, was no different: 'If some race of quadrumanous animals, especially one of the most perfect of them, were to lose by force of circumstances the habit of climbing trees and grasping the branches with its feet in the same way as with its hands . . . and if the individuals of this race were forced for a series of generations to use their feet only for walking and to give up using their hands like feet, there is no doubt . . . that these quadrumanous animals would at length by transformed into bimanous animals.' This new situation would lead to a more intensive development of its senses, and would result in this race 'obtaining mastery over others through the higher perfection of its faculties'.

Lamarck's theory did not, indeed could not, become established. This was not really because of the fallacy of his theory that acquired characteristics are inherited, which in fact only came to be opposed with any force in Darwin's time; it was chiefly because someone greater came upon the scene – a twenty-five-year-old scientist who categorically rejected the theory of evolution.

The Jardin du Roi, later renamed the Jardin des Plantes, the famous research institute in Paris, where both Buffon and Lamarck worked

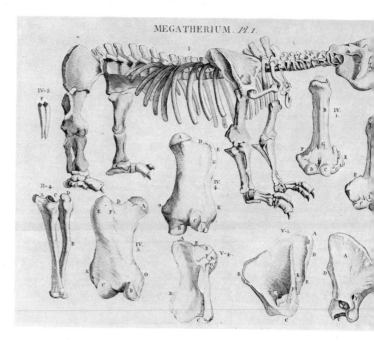

MEGATHERIUM. *Pl. 1.*

Georges Cuvier dominated the study of prehistory until Darwin came on the scene. Right: his reconstruction drawing of the South American giant sloth, 'megatherium'

'Is Cuvier not the greatest poet of our century? This immortal naturalist has reconstructed past worlds from a few bleached bones; has rebuilt cities, like Cadmus, with monsters' teeth; has animated forests with all the secrets of zoology gleaned from a piece of coal. . . . He treats figures like a poet: his zero set against a seven is sublime. He can call up nothingness before you without the phrases of a charlatan. He searches a lump of gypsum, finds an impression in it, says to you, "Behold!" All at once the marble comes to life, the dead live, and a world is laid open before you.'

This hymn of praise, sung by Cuvier's great contemporary Balzac in his novel *Le Peau de Chagrin*, is in fact a historical document. The impression it gives us of the power of Georges Cuvier's influence upon the early nineteenth century is far better than any historical evidence. Cuvier lived from 1769 to 1832. During this period fossils came to be accepted as a serious subject for research and were no longer dismissed as collectors' pieces and mere curiosities. Zoology, geology and palaeontology became established scientific disciplines.

Cuvier laid the foundations for a practical chronology of past periods and the creatures who had lived in them. Fossils were no longer seen as evidence of the sportiveness of nature, or relics of the biblical Flood; they were 'médailles de création', commemorative medals of creation, fossilized circumstantial evidence of a prehistoric life which had existed aeons ago. The only creature which was not part of Cuvier's prehistoric world was the being which reckoned itself supreme on earth – man. He, according to Cuvier, was the product of recent history, a creature without a past, divided from the rest of the animal world by a great gulf and by a series of catastrophes: 'L'homme fossile n'existe pas.'

This was a fateful utterance, an error of judgment made by a seemingly infallible authority. When Lamarck died in 1829, poor, blind and forgotten, Cuvier had become the most celebrated natural scientist in the world. His maxim: 'There is no such thing as fossil man' became elevated into an established doctrine.

That stone weapons and tools from seemingly very ancient times could be found in the earth was then common knowledge. Nevertheless, this did not seem to belie Cuvier's theory. Farmers ploughed them up in their fields, labourers discovered them while building houses or ditches, but no one paid them much attention. They were nothing in comparison with the products of classical antiquity, which the archaeologists of the time treated with such enthusiasm and awe. It was assumed they were the axes, arrowheads and spearheads of the ancient Teutons, Celts, Illyrians, Scythians and other 'barbaric' peoples.

But in 1819 a very novel kind of museum of antiquity was opened in Copenhagen. Here ancient finds were not simply displayed as evidence of wild or semi-wild races of people, but were described by the founder of the museum as belonging quite definitely to three distinct cultural epochs: the Stone Age, the Bronze Age and the Iron Age. This man, a Dane by the name of Christian Jürgensen Thomsen, merchant and amateur naturalist, is today honoured as the 'Father of Archæology'.

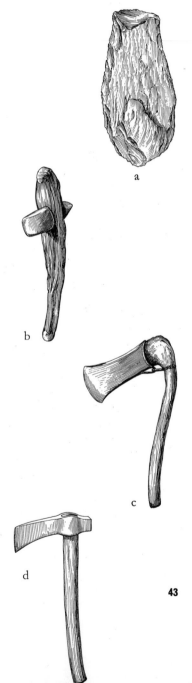

The development of the axe: a) Old Stone Age hand-axe, b) polished stone axe-head in a wooden haft, c) socketed bronze axe-head on wooden haft, d) Iron axe-head on reconstructed shaft (after Oakley)

a

b

c

d

43

Thomsen owes his place in history to two quite new ideas. First, he classified his prehistoric finds as the products of different 'Ages', considering the people of those ancient times to belong not merely to a single epoch preceding classical antiquity but to very different periods, the oldest of them dating back to exceedingly distant times. Secondly, he called these individual cultural epochs after the material out of which the tools typical of each period had been fashioned – after stone, bronze and iron. This Three Age classification, still broadly valid, was the start of the scientific study of prehistoric man and his cultural development.

Christian Jürgensen Thomsen (in top hat) introduced the terms 'Stone Age', 'Bronze Age' and 'Iron Age'. His museum of antiquities in Copenhagen soon became very popular

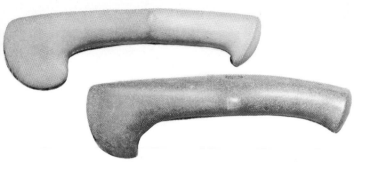

Polished blades of Stone Age battle-axes. Below: double-headed polished flint axe-blade in a reconstructed shaft. The hewn surface of the log of wood shows how cleanly Stone Age man could work with his 'primitive' tools

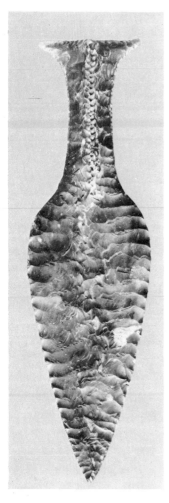

A Late Stone Age flint dagger

Perforated mussel shells dating to the Stone Age, from Ofnet in Bavaria

The apparently straightforward and practical term 'Stone Age', however, turned out to be an exceedingly problematic one. Jens Jakob Asmussen Worsaae, a pupil of the great Thomsen, betook himself to England in 1846 with written credentials from the Danish king to study the 'antiquities' of the British Isles. Here he found much which cast doubt on the validity of Thomsen's Three Age system. The implements which Thomsen had described as belonging to the Stone Age could by no means be called primitive. According to him, the only difference between the tools of this period – daggers, knives, axes and saws, made of flint, jade, nephrite or obsidian – and those of the later periods was that they had been fashioned out of stone rather than metal. But in England and France a very different sort of stone artefact was unearthed: crude, unwieldy, wedge shapes whose usefulness as an implement was barely apparent, and knives, scrapers and scorers simply knocked out of a piece of flint. They were found in the gravel beds of rivers, in clay and gravel ditches and in the sediment of subterranean caves, often scattered among the bones of 'antediluvial' animals.

If these primitive wedge-shaped tools of stone were to be accepted as the implements of an early race of man then there must have been two Stone Ages quite clearly distinguishable from one another, an Old Stone Age and a New Stone Age, as the English prehistorian Sir John Lubbock later named them. Was the difference between them merely that the Old Stone Age tools were crude and unpolished, whereas those of the later period had been made with great precision, were burnished, and sometimes even shaped? Or should the Old Stone Age be identified with that far distant epoch which had once been linked with the biblical deluge and was therefore called the Diluvium, the time of the Flood? The discovery of an Old Stone Age reawakened interest in the search for 'antediluvial' man, that is, prehistoric man.

In the Late Stone Age, about 4000 BC, the first farmers settled in Central Europe and the Danube valley

In two caves near Liége P.C. Schmerling (above) found the skulls of prehistoric men (bottom)

'The situation in which these weapons were found may tempt us to refer them to a very remote period indeed; even beyond that of the present world.' This comes from a letter sent by one John Hookham Frere to the Society of Antiquaries in 1791. He enclosed some stone artefacts. Frere, a distinguished diplomat and writer, had found these ancient implements in a clay ditch near Hoxne in Suffolk, together with the bones of 'diluvial' animals.

This was not the first evidence for the existence of prehistoric man to have come to light. There had been reports of fossilized human bones in a cavern on the Island of Cerigo in Italy, in a variety of caves in southern France, at Köstritz and Bilzingsleben in Thuringia, and near Lahr in Baden. But the most important discovery was made in 1830 by the Belgian doctor of medicine P.C. Schmerling, in the caves of Engis and Engihoul near Liége. He found seven human skulls and a number of human bones, together with primitive flint artefacts and remains of mammoths and other Ice Age creatures. Schmerling's report runs as follows: 'Whatever conclusion we like to reach about the origin of mankind, I for one am convinced that a skull of this kind belongs to a person of limited intellectual faculties, that is, a man belonging to a lower level of civilization. . . . There

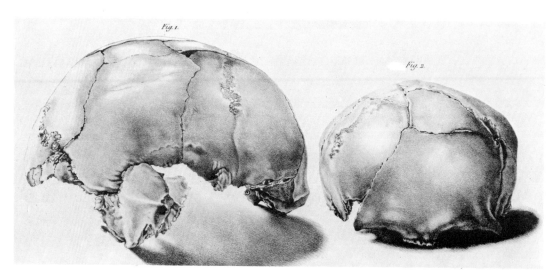

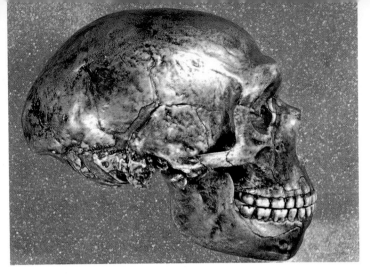

Top: *This skull of a Neandertaler belongs to an almost complete skeleton found in a cave near La Chapelle-aux-Saints, France* Bottom: *Reconstruction painting by Burian of a Neandertal living site*

can be no doubt that these human bones came to lie in the earth at precisely the same time and for precisely the same reason as the bones of other extinct animals.'

Anthropology, dominated by Cuvier's theories, had simply ignored such finds. But now, in England, discoveries were accumulating rapidly.

The 'Red Lady of Paviland' was responsible for triggering off a fresh search for antediluvian man similar to the search made in Scheuchzer's time. A well-preserved skeleton (without its skull) of what has since proved to be a prehistoric man, stained a deep red from its bedding of red ochre, was discovered in the so-called 'Goat's Hole' on the Welsh coast.

It was William Buckland, an authority on science and a faithful adherent to orthodox biblical belief, who made the discovery. In *Reliquiae Diluvianae*, which he published in 1823, he described five other places in the British Isles containing evidence of prehistoric times, one of them Kent's Cavern, near Torquay in Devonshire. Here, under a solid layer of stalagmite, a Roman Catholic priest, John McEnery, found a number of flint tools, together with the bones of cave bears, woolly rhinoceroses and other extinct animals. Buckland reckoned that his *Reliquiae Diluvianae*, 'by affording the strongest evidence of a universal deluge, leads us to hope that it will no longer be asserted as it has been by high authorities, that geology supplies no proofs of an event in the reality of which the truth of the Mosaic records is so materially involved.'

While the pious Dean Buckland was only out to prove that the Flood had really occurred, a compatriot of his was searching for evidence that man was older than the Bible would have us believe. William Pengelly, son of a humble workman, had left his village school when he was only twelve, spending the next four years at sea as ship's boy and cook's mate. But his hunger for knowledge was so intense that he gave up his seafaring existence. By intensive private study he gained enough knowledge of the natural sciences to become a teacher at a school for craftsmen. Later he

Opposite: Burian's impression of a Neandertaler

Longitudinal section of 'Goat's Hole', Paviland, in Wales, from William Buckland's 'Reliquiae Diluvianae', 1823

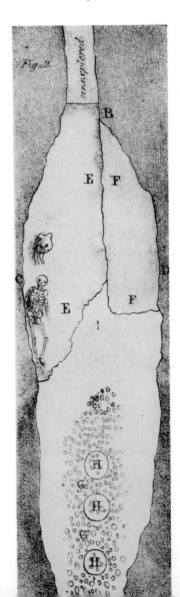

Buckland giving a lecture on geology in the Ashmolean Museum, Oxford

William Pengelly, one of the first to recognize Old Stone Age tools

founded his own school in Torquay, not far from Kent's Cavern. Using this as his base, he made a thorough investigation of all the caves in the neighbourhood, determined to discover whether the stone implements to be found in these caves were really contemporaneous with the bones of extinct animals.

Assisted by five distinguished geologists, Pengelly first systematically investigated Kent's Cavern, then Windmill Hill above Brixham harbour. The layer of stalagmite covering the floor of the cave was hard and even; hence it had definitely not been tampered with. And just beneath this covering lay the bones of mammoths, woolly rhinoceroses, reindeer, cave lions, cave bears and hyenas, all side by side, and amongst these flint implements which had quite clearly been fashioned by human hand. Thus men must have lived alongside these prehistoric animals. Cuvier's theory was completely routed: fossil man existed after all.

An official verdict was reached in 1859, when Pengelly's work was finally recognized by the scientific societies in England. In the same year Charles Darwin published his theory of evolution; and in this year too a Frenchman who had for close on thirty years fought in vain for 'Adam's ancestors' achieved recognition at last. This was Jacques Boucher de Crèvecœur de Perthes, one-time diplomat and writer, and, like Pengelly, a gifted amateur archaeologist.

After a varied diplomatic and literary career, Boucher de Perthes was appointed customs officer in Abbeville in north-western France at the age of thirty-seven. A year after he had taken up his post he came across his first lumps of flint in a gravel pit in the Somme valley. These stones, some sharp-edged, others pointed, were later to make his name immortal. Noticing that they lay in the stratum of earth which geologists called 'antediluvial', he persuaded the workmen in the stone quarries of Moulin Quignon to collect the bones of mammoths, prehistoric horses and other creatures of the Diluvium for him from the same deposit.

Jacques Boucher de Crèvecoeur de Perthes fought for years to prove that the lumps of flint he had found were Old Stone Age tools

In 1832 he picked up a stone which baffled him. The lump looked as if it had been artificially shaped: one end fitted a man's hand exactly, and the other was pointed, like a chisel or some hand-axe. Looking through his collection, he discovered many similarly shaped stones. But it was only in 1838, after he had collected a large number of these tool-like lumps of flint, that he dared mention his finds in Abbeville. A year later he delivered a lecture on the subject in Paris. His flints, he declared, were thousands of years older than all previous archaeological discoveries. Despite their imperfections they were no less evidence of the existence of early man than an entire museum.

The experts listened politely, but were unconvinced, declaring the flints to be worthless, one of nature's accidents. But Boucher de Perthes would not be dissuaded. He carried on collecting, and in 1847 published a three-volume work on his discoveries. Some experts, Worsaae among them, showed much interest; but in France itself Boucher de Perthes found his theory sharply challenged. He complained bitterly about his critics: 'I now had my evidence of prehistoric times, but I had it for myself alone. . . . Others were frightened, unwilling to be associated with such heresy. The weapon they used against me was far more powerful than any opposition, criticism, satire, or even persecution could have been – the weapon of disdainful silence. No one discussed the facts I had to present, they did not even trouble to refute them – they just ignored them.'

In 1850 his colleagues resolved to rid themselves of this awkward customer. One of his sharpest opponents, a certain Dr Rigollot of Amiens, decided to make his own excavations in the gravel-pits and stone quarries of the Somme valley so that he could dispose of the crackpot of Abbeville's theories once and for all. Yet the result of his investigations was precisely the opposite: they merely revealed how right Boucher de Perthes had been. The disputed wedges of flint had unquestionably been fashioned by human hand. Not only that: they lay in a deeper stratum than the English finds which had caused such a stir, were also more primitive in appearance, and hence must date back even further. The Old Stone Age culture which Boucher de Perthes had been so lucky to come across has been named the Acheulean, after Saint-Acheul where the flints were first found. They belong, we now know, to the most ancient of European Stone Age cultures, and are generally attributed to the Presapiens-Preneandertal group, that is, the early ancestors of late Ice Age man.

Rigollot now gave unqualified support to Boucher de Perthes' theory. Now too – that is, between 1835 and 1854 – leading geologists from England began to travel to the Somme valley to inspect the finds. The excavations near Abbeville and Saint-Acheul became regular places of pilgrimage for British and French geologists. Their verdict was summed up enthusiastically by the English geologist Sir Andrew Ramsay, who wrote that for more than twenty years he and other men of his profession had regularly handled natural and artificially shaped stones. But for him the controversial axes of Amiens and Abbeville seemed 'just as definite evidence of human activity as the knives of Sheffield'.

Some of the controversial flint axes which Boucher de Perthes illustrated in his 'Antiquités celtiques et antédiluviennes'

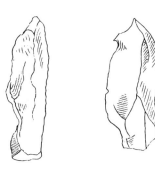

'I came, I saw, I was conquered', admitted one of the most distinguished English scientists on his return from Abbeville. He was speaking to a group of British naturalists gathered expressly to hear his verdict on the Acheul flints. This was Sir Charles Lyell, the greatest geologist of his day, 'The Chancellor of the Exchequer among scientists', as Darwin called him. Lyell's declaration caused a tremendous stir and was considered a revolutionary verdict, as indeed it was. For it brought about the fall of a god – the fall of the Titan, Cuvier.

Lyell, eldest son of a rich landowner, gave up studying law at an early age in order to devote himself to geological research. From 1830 to 1833 he wrote the most important work on geology since the Renaissance, *The Principles of Geology*, which appeared in three volumes. In this he showed how the earth had gradually and imperceptibly changed, affected by those same geological phenomena we still observe happening today. He accepted the theory of an Ice Age, as put forward by Swiss scientists, according to which the alleged time of 'catastrophic flood' was not one of inundation but of glaciation. Also, he did not reckon the age of the earth in thousands or tens of thousands of years, but based his ideas on 'immense periods of the past'.

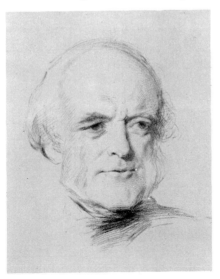

Sir Charles Lyell, founder of modern geology

Attempts were made during his lifetime by the physicist Sir William Thomson, later Lord Kelvin, to determine the time between the earth's original state of red heat and the present day. Although Kelvin was a deeply religious man, he was the first scientist to calculate coolly in millions, proclaiming the earth must be at least 24 million years old. Today, on the basis of the 'uranium clock', which measures the decay of radioactive substances in stone, and on the basis of other methods too, we reckon the earth to be some 5,000 million years old.

Lyell's theory of a gradual process of change in the earth's surface could be expected to result in the idea of a similar slow and gradual evolution of life and man. But it took twenty-five years for the great English geologist to overcome his scruples and accept that this was so. Once in a letter to his friend Darwin he admitted, 'I have long seen most clearly that if any concession is made, all that you claim in your concluding pages will follow.'

It was Lyell's hesitation which caused Schmerling, McEnery and other pioneers in the study of prehistory to strive so long in vain for recognition. Nevertheless, Lyell

Panoramic sketch of the glaciers near Monte Rosa from 'Etudes sur les glaciers', published in 1840 by Louis Agassiz, whose glacier theory provided the basis for our present understanding of the Ice Age

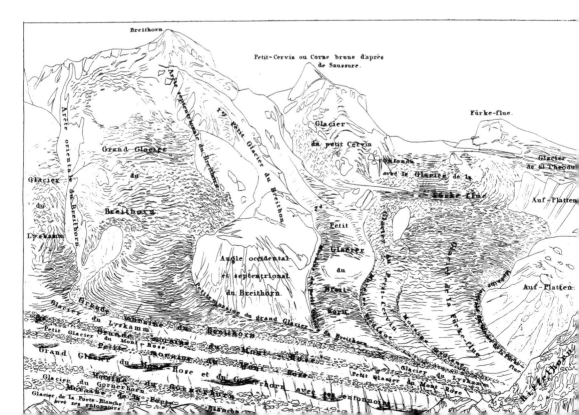

visited every place where prehistoric finds had been reported. The evidence produced by these discoveries gradually became more and more overwhelming, till, in the historic year of 1859, Lyell announced his conviction that the
discoveries of Boucher de Perthes and William Pengelly
must be authentic after all. In his speech to the Geological
Society of Aberdeen he admitted: 'The facts recently
brought to light . . . must, I think, have prepared you to
admit that scepticism in regard to the cave evidence in
favour of the antiquity of man had previously been pushed
to the extreme.'

A year later Lyell visited the most interesting and also the
most controversial prehistoric site: the Neandertal, or
Neander valley, which is situated not far from the Rhenish
city of Düsseldorf.

The pious Joachim Neumann, clergyman and organist,
had written many hymns under the pseudonym Neander.
A nature-lover, he had often visited the wild and romantic
glen near Mettmann, through which a stream, the Düssel-
bach, flows; and because he had composed many of his
most popular hymns there the valley came to be called the
Neandertal, which in turn gave its name to the most famous
of all fossil men, and is indeed almost synonymous with
prehistoric man himself. However, about the middle of the
nineteenth century the valley began to lose some of its
natural beauty, when quarrying was begun and a narrow-
gauge railway for transporting the rock was built along
one side of it. In 1856, as a result of those operations,
two caves, known as the Feldhof grottoes, were exposed
some sixty feet above the valley level.

As the workmen were removing the six-foot layer of
earth from the smaller of the two caves they came across
some decayed bones. Herr Beckershoff, part-owner in the
business, glanced at them casually. He took them to be the
remains of a cave bear and instructed the workmen to search
the cleared debris for further bones. As a result Beckershoff
found himself in the possession of an unusually large skull-
cap and various other parts of a human skeleton. He kept

them a few weeks in his home, but then handed them over to Johann Carl Fuhlrott, the president of the local naturalists' society in Elberfeld and a teacher by profession.

The greatest credit must go to Fuhlrott for evaluating the Neandertal discovery more or less correctly. The bones could not have been those of a modern man; therefore they must belong to early man, or – as Fuhlrott put it – to 'a typical very ancient individual of the human race'.

The Bonn anatomist, Hermann Schaaffhausen, subjected the bones to a detailed analysis and agreed with Fuhlrott. Schaaffhausen was particularly impressed by the 'prominent forehead which in fact is similar to that of a great ape'. After some hesitation and a thorough comparison of the bones with those of other ancient and modern human skulls, he came to the conclusion that the 'remarkable human remains from the Neander valley can be considered the oldest evidence we have of the early inhabitants of Europe'.

During the following months and years Fuhlrott and Schaaffhausen did all they could to convince their naturalist colleagues, but in vain. The experts produced a wide variety of very contradictory opinions: the Neandertal man was 'undoubtedly a Celt', or an 'ancient Dutchman', or 'a poor hydrocephalic idiot who had lived like an animal in the forest', a 'hermit', or even 'a wild cannibal who had somehow been transported to Europe'. The scholarly verdict reached by a fellow anatomist of Schaaffhausen's in Bonn, Robert Mayer, was that the bones could only be those of a Cossack of the Russian army, who had fallen when the army crossed the Rhine in January 1814 under General Tchernitchev. The femurs of the Neandertal man were bent inwards, and according to Mayer were 'the typical thigh bones of a man who has ridden horses constantly since his youth'.

The most vigorous refutation of the contention that these bones belonged to prehistoric man came from the great medical authority of the day, Rudolf Virchow, who enjoyed much the same esteem in his time as had Cuvier in

This skull-cap started the ball rolling: it belongs to Fuhlrott's violently contested Neandertal man

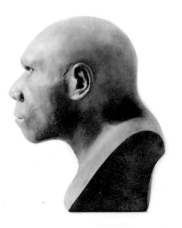

his. When Fuhlrott delivered a lecture about his discovery at a scientific convention in Kassel in 1857, Virchow answered him with a brilliant, splendidly ironic speech, culminating in the verdict that the skull was not prehistoric but pathological. The curved leg-bones were the result of rickets, he declared, the knots of bone above the eyes had been caused by damage to the skull, and other special features of the skeleton were the result of *arthritis deformans*. Despite these grave physical defects, the wretched individual had reached a ripe old age. And as this could only have been possible within an advanced social community, and as such a community cannot have existed during the Ice Age, all hypotheses that these bones were ancient had to be rejected.

Thirty years later, when Darwin's theory had become the focus of scientific attention, Virchow brought up the Neandertal discovery again and repeated his emphatic denial that the bones could be prehistoric, and also that man's origin was animal. 'The idea that men arose from animals is entirely unacceptable in my view, for if such transitional men had lived there would be evidence of it, and such evidence does not exist. The creature preliminary to man has just not been found.'

Lyell and other English scientists took a more positive line. William King, the anatomist, went so far as to endow the type of man discovered in the Neander valley with the scientific name of *Homo neanderthalensis*; and Darwin's protagonist, Thomas Henry Huxley, termed it 'the most ape-like human skull I have ever seen'. Nevertheless, the greater part of the scientific world remained sceptically reticent, and the Elberfeld naturalists' society felt the need to disassociate itself from its President. Although Fuhlrott was allowed to publish an article in the society's journal, the editor was moved to append his own comment: 'We feel obliged to state that we do not share the opinions expressed here.'

When Fuhlrott died in 1877 the Neander valley discovery had been almost forgotten. Darwin hardly mentions it in his work. It was only after the turn of the century that the skeleton found in the Feldhof grotto suddenly became the focus of interest once more, as we shall see.

'At last gleams of light have come, and I am almost convinced . . . that species are not (it is like confessing a murder) immutable,' wrote the thirty-five-year-old Darwin to Sir Joseph Dalton Hooker, Director of Kew Gardens. These are over-cautious words, uttered by a hesitant man who was constantly assailed by doubt and whose very nature recoiled at the idea of himself as a scientific revolutionary. One of his biographers, William Irvine, has described his character thus: 'He read so slowly, wrote so slowly, even thought so slowly, that he always felt desperately behindhand, like a tortoise concentrating every energy on the next step, as he creeps in frantic haste toward impossible horizons. . . . He had a gentleman's fear of being conspicuous, an invalid's sensitiveness to the idle curiosity of crowds. How could he come before the nation with

The thirty-year-old Charles Darwin, three years after his epoch-making journey round the world

theories that were not only a scandal to orthodox biology, but a blasphemy against religion and Victorian decency?'

As he worked on his book, *The Origin of Species*, evolving a doctrine which was to shatter all prevailing ideas of the world and man, Darwin said of himself on the discovery of some slight error he had made: 'I am the most miserable, bemuddled, stupid dog in all England, and am ready to cry with vexation at my blindness and presumption!' In a letter to Charles Lyell about the same period he admitted, 'I have asked myself whether I may not have devoted my life to a fantasy'.

It is fascinating to trace the development of Darwin's theory of selection from his own writings and those of his contemporaries, and to follow the course of the ensuing controversy and the battles which were fought over this new scientific doctrine. As far as the theory of evolution is concerned, Darwin had an inherited interest. He came from a family of medical and scientific men. His grandfather, the eighteenth-century doctor, naturalist and poet, Erasmus Darwin, had written many books on natural science as well as didactic poems, and had expressed a view of the development of life which was much along the lines of the theory Lamarck was later to expound. Oddly enough Darwin was entirely uninfluenced by these ideas. He once wrote: 'It is curious how largely my grandfather, Dr Erasmus Darwin, anticipated the views and erroneous grounds of opinions of Lamarck.' Darwin, having twice read through Lamarck's *Zoological Philosophy* thoroughly, pronounced it 'a wretched book'. It was natural that Darwin, who 'worked on true Baconian principles', collecting his facts by observation and experience, should feel a deeply rooted dislike of all philosophical speculation.

Darwin states repeatedly in his diaries and writings that his theory of the origin of species had arisen out of his first journey round the world, which took place between 1831 and 1836. It is odd to think that Darwin very nearly did not make the trip, that he very nearly did not take part in that survey carried out by H.M.S. *Beagle* which was to have such far-reaching and historical repercussions – merely because of the shape of his nose. The Captain of the *Beagle*, Robert Fitzroy, was in Darwin's words 'an ardent disciple of Lavater and was convinced he could judge a man's character by the outline of his features'. When the twenty-two-year-old Darwin was recommended for the post of scientific assistant by his University teacher, the botanist and zoologist John Stevens Henslow, Fitzroy reckoned the young Cambridge student's nose indicated an indecisive character. And in this he was not entirely wrong. It was only after considerable hesitation that Captain Fitzroy decided to take him on. He could not have foreseen the outcome of this decision. Thirty years later, when Darwin's doctrine was proclaimed in Oxford, he gestured with the Bible in token of his distaste for his one-time fellow traveller's arguments.

Robert Fitzroy as Vice-Admiral.

Marine lizards of the Galapagos Islands. The animals living on this island first set Darwin's mind on the trail of his theory of evolution

The changing appearance of three Fuegians. How they looked during a stay in England in 1832 (top and bottom), and after returning to their native surroundings (centre). Drawings by Captain Fitzroy from his book 'Narrative of the Surveying Voyages of H.M.S. "Adventure" and "Beagle", 1838'

Las Encantadas, the enchanted isles, is how the Spanish designated the Galapagos Islands when they first encountered them. Volcanic in origin, they are situated a thousand miles west of Ecuador in the Pacific Ocean. The scientific expedition of the *Beagle* dropped anchor here in October 1835. On the islands Darwin found a fascinating animal world, for while it did indeed have some affiliations with the corresponding animals of South America, it was not at all the same: giant tortoises, marine lizards, blunt-snouted iguanas, and notably finches, later to be known as Darwin's finches and be the subject of considerable scientific discussion. The ancestors of these creatures had obviously strayed to the 'enchanted' islands from the mainland of South America, but in the course of time each species had given rise to new types and forms. The impressions Darwin gained on the islands led him to the conclusion that forms could not be immutable, that some process of evolution must in fact take place. But all we have in his diary of this idea is one short sentence: 'With this we appear to have come somewhat closer in space and time to that great fact – the mystery of mysteries – the appearance of new beings on earth.'

'No fact in the long history of the world is so startling as the wide and repeated exterminations of its inhabitants.' This statement reads like a comment made by one of Cuvier's followers, believing in universal catastrophes and the complete extermination of the antediluvial worlds, not like that of a man believing in the gradual evolution of species. In fact this remark was made by Darwin on the voyage of the *Beagle* when, with Captain Fitzroy's permission, he left the ship for a brief while to dig for fossils in the Tertiary strata of Patagonia. He discovered the remains of toxodons, giant sloths and other strange-looking mammals which according to the Cuvier school of thought had all perished in some kind of cataclysm. 'The mind is at first irresistibly hurried into the belief of some great catastrophe,' Darwin notes, 'but thus to destroy animals, both large and small, in Southern Patagonia, in Brazil, on the Cordillera of

Peru, in North America up to Behring's Straits, we must shake the entire framework of the globe. An examination, moreover, of the geology of La Plata and Patagonia, leads to the belief that all the features of the land result from slow and gradual changes.'

Thus Darwin adopted Lyell's theory of a gradual change taking place on the earth and applied it to organic life as well. He saw that many fossils bore a marked similarity to living species and must in some way be related to them. Had these prehistoric animals not merely become extinct? Should we not rather consider them the genuine physical forbears of present-day forms of life? Darwin came to the following conclusion: 'This wonderful relationship in the same continent between the dead and the living will, I do not doubt, hereafter throw more light on the appearance of organic beings on our earth, and their disappearance from it, than any other class of facts.'

'The subject haunted me', Darwin subsequently wrote in his Autobiography. He sought proof of the change which species undergo, starting from a fact overlooked by all his predecessors, which is that man is able to create new domestic species by selective breeding, controlled selection. Did nature too practise some sort of selection? 'I soon perceived', he writes, 'that selection was the keystone of man's success in making useful races of animals and plants. But how selection could be applied to organisms living in a state of nature remained for some time a mystery to me.' In October 1838 he happened to read a controversial book entitled *An Essay on the Principles of Population*. Its recently deceased author, Thomas Robert Malthus, a clergyman, had pointed out the dangers of overpopulating the earth. Nature, explained Malthus, prevented the overproduction of living creatures by making them fight for their existence; but man, to whom this no longer applied, must impose his own limits upon the rate of his reproduction, so that future generations should not want, starve and finally die.

It was in Malthus' book that Darwin first found the phrase, the 'struggle for life'. Malthus' recognition of the fact that nature, in her wisdom, regulates the number of animals was more important to Darwin than his rather unrealistic advice that human reproduction should be restricted by sexual abstinence. 'Being well prepared to appreciate the struggle for existence which everywhere goes on from long-continued observation of the habits of animals and plants, it at once struck me that under these circumstances favourable variations would tend to be preserved, and unfavourable ones to be destroyed. The result of this would be the formation of a new species. Here, then, I had at last got a theory by which to work; but I was so anxious to avoid prejudice that I determined not for some time to write even the briefest sketch of it.'

The study at Down House, where Darwin formulated his new theories. The picture dates from the year of his death, 1882

In 1871 this caricature of Charles Darwin by 'Ape' appeared in 'Vanity Fair'

'*Vanity Fair*' *also published this caricature by* '*Ape*' *of Thomas Henry Huxley in 1871*

But at the same time, as Darwin himself later admitted, he had 'overlooked a problem of great importance'. His theory of a natural selection in the struggle for existence could only be valid if the offspring of a pair were not all exactly the same but tended to diverge considerably in character. And this could only be proved by a series of tedious experiments. In 1842 Darwin retired with his family to Down House in the village of Downe, Kent and worked there on his own, in complete seclusion. He had contracted a severe disease on his journey round the world, and his friends and colleagues were not surprised to hear little of him for the next fifteen years. Only a few trusted friends had any idea that the invalid, living in almost eccentric seclusion, was breeding different species of animals and plants, substituting artificial selective breeding for natural selection in an attempt to discover the mechanism of selection.

Darwin did not yet occupy himself with the problem of man. In a letter to Alfred Russel Wallace, another proponent of the theory of evolution, he declared, 'You ask me whether I draw man into the discussion. I think I shall avoid the whole subject as so surrounded with prejudices; though I fully admit that it is the highest and most interesting problem for the naturalist.' And in his epoch-making book, *The Origin of Species by Means of Natural Selection*, which was to evoke probably the greatest controversy in the history of biological research, the only mention of man is in one concluding sentence: 'Light will be thrown on the origin of man and his history.'

Alfred Russel Wallace, one of Darwin's protagonists, in the Borneo Jungle

'What is the question now placed before society with a glib assurance the most astounding? The question is this: Is man an ape or an angel? My lord, I am on the side of the angels.'

In 1864, five years after the publication of Darwin's book, which despite the cautious phraseology of its author had had such repercussions upon every aspect of cultural, intellectual and social life, no less a person than the British Prime Minister, Benjamin Disraeli, felt moved to make clear his own position with these words. Other verdicts were sharper still. Darwin's old geology professor, Adam Sedgwick, on reading the book, declared there was nothing in it on morals and metaphysics, and that 'a man who denies this is deep in the mire of folly'.

But even Darwin's most well-meaning friends and supporters could at first make little of his theory of selection. The great Sir Charles Lyell, whose book on geology had had such a considerable influence on Darwin, and who himself believed that people had existed during the Ice Age, resisted the radical rethinking Darwin's theory obliged him to make. Hooker summed up Lyell's reaction to *The Origin of Species* with the words, 'Lyell . . . is somewhat staggered.' What caused Lyell to hesitate was that 'ticklish question', the origin of man. Despite his acceptance of man's great age, he was obviously fearful of acknowledging the final consequences of it. 'It is a good joke,' wrote Darwin to Hooker of Lyell's hesitation. 'He used always to caution me to slip over man.' Even the naturalist and traveller Alfred Russel Wallace, who had drawn up a theory of evolution along the lines of Darwin's theory as early as 1858, was not prepared to admit that God had no hand in the creation of mankind. 'Our brains are made by God and our lungs by natural selection.' The eminent prehistorian, Hugh Falconer, a colleague of Lyell's, even accused Darwin of corrupting such admirable scholars as Lyell and Hooker, declaring, 'You will do more harm than any ten Naturalists will do good!' Newspapers and journals printed caricatures of Darwin as an ape, and an indignant reader from Wales described the head of the man who had created the theory of selection as that of 'an old ape with a hairy face and a thick skull'.

Nevertheless, many of Darwin's critics, Lyell among them, ended up as his staunch defenders, convinced of the truth of his contentions. Four years after *The Origin of Species* Lyell brought out *The Antiquity of Man*, which went far in its support of the theory of evolution, although, in the verdict of one biographer, the book 'begins as a geological treatise and ends as an essay on liberal theology'. But, although many scientists were at first uncertain what to make

of it, one naturalist of some standing was instantly converted against all expectation to the 'heresies' of the selection theory. 'No work on Natural History Science I have met has made so great an impression upon me,' he wrote, on reading the book, adding with a sigh, 'How extremely stupid of me not to have thought of that myself!' This was the anatomist, physicist and marine biologist, Thomas Henry Huxley.

67

16. What is Domestication — changed conditions

19 — whether some slight variation must be admitted.

20 Changes in the individual

21 Congenital variation

22 Hereditariness — why are they inherited & not constant. — Mutilation effect a mind. not heredy
     p 25 antiquity p. 27 Prenatal affected & sex. p 27 variation appearing at same age

p. 29. Causes of variation — Direct causes p. 31 Habit p. 32 variation from organisation
     p 34 indirect effects, making organs plastic.

p. 35 Laws regulating variation — Balancement — misplaced features. — p 37 effect of homologous
     38 mechanical relations — arrested p. 39 & 40 cohesion — multiple organs easy in number
     p 41 variation analogous to the species — 41 condition of youth & variation
     — Correlation, age colour & constitution.

p. 43 Effect of crossing in obliterating & fixing races; selection required

p 47 Selection, produces effect of adding up small changes. p 50 Breed are true p 51 antiquity
     selection : p 54 unconscious selection : individual & natural selection : p 56 acclimatisation
     of plants. p 56 Selection for & against. & selection; non-varying: facility in securing object
     64 effects of selection on natural results; 67 effects of selection, physical & spiritual compared.

(Chap. II Var. under Domest. (continued)

3. general argument whether plant is much altered of domest., or not to be recognised: steps
     in cultivation of Cabbage — why certain variation have not produced useful products.

10 The Cabbage

13 Dog

18 —— changes within historical time
20 Cat
22. Horse

26 Pig

30 Cattle

35 Sheep & Goats

40 Rabbit

43 Fowl

46 Ducks

50 Pigeons	52 Pouter	54 Fantail	56 Jacobin
57 Tumbler	59 Turbit	60 Barb	61 Carrier
62 Runts	65 Dovecote &c	68 improperly kept Breeds	
69 Amount of imputed differences			
70 Individual variability			

'**Darwin's bulldog**' was the nickname applied to Huxley within months of publication of *The Origin of Species*. He referred to himself somewhat more elegantly as 'Darwin's agent'. Darwin himself, as befitted a man of his retiring temperament, kept well out of the controversy aroused by his book, feeling that those in the battle 'are growling at one another in a manner which is far from gentlemanlike'. But Huxley plunged right into the thick of it: 'As to the curs which will bark and yelp ... you must recollect that some of your friends at any rate are endowed with an amount of combativeness that will stand you in good stead. I am sharpening up my claws and beak in readiness.'

Huxley kept his promise. Moreover, he had no compunction about tackling the problem of human origin. When on 30 June, 1860 eminent scholars and other persons in public

Opposite: The first handwritten page of Darwin's scheme for 'The Origin of Species'

Huxley's opponent Bishop Samuel Wilberforce caricatured in 'Vanity Fair' after the Oxford trouncing of 1860

life met in Oxford for a meeting of the British Association to discuss Darwin's claims, Huxley delivered a brilliant lecture about the theory of selection, in which he made a particular point of tackling the problem of the origin of mankind and the physical affinity between the brain of an ape and that of a man. The speaker for the opposition was Bishop Samuel Wilberforce, whose unctuous manner had earned him the nickname Soapy Sam. Wilberforce's grounding was obviously far from thorough, for he was reduced to trying to ridicule Huxley, asking him whether he truly believed he was descended from an ape, for then 'it would be interesting to know whether the ape in question was on your grandfather's or your grandmother's side.'

As Huxley was later to write in his memoirs, he whispered to his neighbour, 'the Lord hath delivered him into my hands,' and answered Wilberforce quietly: 'If you ask me whether I would rather have a miserable ape for a grandfather or a man highly endowed by nature and possessed of great means of influence and yet who employs those faculties and that influence for the mere purpose of introducing ridicule into a grave scientific discussion, then I unhesitatingly affirm my preference for the ape.'

Although a lady fainted with horror and Darwin's onetime captain, Robert Fitzroy, made as if to throw the Bible at Huxley, the confrontation ended in a victory for the 'bulldog'. This had taught Huxley to speak before a large audience. He later told Hooker that from then on he had gradually overcome his dislike for public speaking by carefully cultivating the art.

'The question of all questions for mankind – the problem which underlies all others, and is more deeply interesting than any other – is the ascertainment of the place which Man occupies in the universe of things. Whence this race has come; what are the limits of our power over nature, and of nature's power over us; to what goal we are tending; are the problems which present themselves anew and with undiminished interest to every man born in the world.'

Thomas Henry Huxley, drawn by his daughter, Marian Collier

This is the crux of Huxley's three lectures of 1863, *Man's Place in Nature*, which appeared shortly afterwards in book form. It was the first thorough study of the racial origin of man. In it, Huxley described the known remains of prehistoric men, including the Neandertal man whose significance was still being disputed, and compared their bones with those of various lower and higher species of ape. In his view the Primates, Linnaeus' 'master animals', made up an unbroken ladder leading 'from the crown and summit of the animal creation down to creatures from which there is but a step, it seems, to the lowest, smallest, and least intelligent of the placental Mammalia. It is as if nature herself had foreseen the arrogance of man, and with Roman severity had provided that his intellect, by its very triumphs, should call into prominence the slaves, admonishing the conquerer that he is but dust.'

Careful anatomical studies and the comparison of brains had led Huxley to a conclusion which at the time was considered highly provocative but is now obvious to every anthropologist and primatologist: 'Whatever system of organs be studied, the comparison of their modifications in the ape series leads to one and the same result – that the structural differences which separate Man from the Gorilla and the Chimpanzee are not as great as those which separate the Gorilla from the lower apes.'

Prehistoric man (Eohomo) rides a prehistoric horse (Eohippus). Thomas Henry Huxley illustrated the theory of evolution with this humorous drawing

'Our ancestor was an animal which breathed water, had a swim bladder, a great swimming tail, an imperfect skull, and undoubtedly was a hermaphrodite,' was Darwin's cautious phrasing of it in a letter to Lyell soon after Huxley had raised the problem of man's origin. Darwin was only prepared to talk of man's far distant ancestors who had belonged to the realm of fishes, and avoided speaking of apes. Huxley's relish for combat, however, made him write to his wife before a lecture, 'By next Friday evening they will all be convinced that they are monkeys!' But Darwin came only gradually, step by step, to approach the problem. True, he greeted Lyell's *Antiquity of Man* with the friendly

Charles Darwin in old age

words, 'What a fine long pedigree you have given the human race!' But he did not draft theories or set up genealogical trees. He collected facts. He started his investigations by studying the distinctive physical characteristics which could lead one to suppose that man must be related to the mammals most similar to him; for instance, the fine covering of hair which covers the human foetus at six months, the wisdom teeth, the appendix, the rudimentary tail indicating our prehistoric ancestors had once lived a forest life, our upright posture which had first made it possible for man to use his hands for other things than mere locomotion, such as making tools. Finally, in February 1867, Darwin wrote to Wallace: 'I have almost resolved to publish a little essay on the origin of mankind, and I still strongly think ... that sexual selection has been the main agent in forming the races of man.'

When *The Descent of Man and Selection in Relation to Sex* was published in 1871 it was again Darwin's cautious and meticulous presentation of the facts which aroused the most indignation. Far from deriving man from any species of ape still living, he made a point of stressing that: 'We must not fall into the error of supposing that the early progenitor of the whole simian stock, including man, was identical with, or even closely resembled, any existing ape or monkey.' In his words the ancestor of man was 'an ancient member of the anthropoid sub-group', a definition which still holds good. His assumption that these prehistoric forefathers of ours had presumably lived in Africa has been made at least probable by the discoveries of recent years.

Darwin was of course perfectly well aware that he would hurt the pride of his contemporaries by these seemingly timid statements, and he makes a fitting apology in his conclusion: 'Man may be excused for feeling some pride in having risen, though not through his own exertions, to the very summit of the organic scale; and the fact of his having thus risen, instead of having been aboriginally placed there, may give him hope for a still higher destiny in the distant future.'

'No sooner does the naturalist discover the resemblance of some higher mammals, such as the ape, to man, than there is a general outcry against the presumptuous audacity that ventures to touch man in his inmost sanctuary. The whole fraternity of philosophers, who have never seen monkeys except in zoological gardens, at once mount the high horse, and appeal to the mind, the soul, the reason, to consciousness, and to all the rest of innate faculties of man, as they are refracted from their own philosophical prisms.'

Karl Vogt, nicknamed 'Affenvogt', 'monkey-keeper', because of his enthusiasm for Darwin's doctrine

73

Every unbiased specialist in the study of apes today will confirm these words of Karl Vogt. In his *Lectures on Man, his Place in Creation and in the History of the Earth*, this witty, rather portly scientist, who had been a zoologist in Giessen but whose radicalism had obliged him to move to Geneva, took up Darwin's theory wholeheartedly. The scientific world soon nicknamed him 'Affenvogt', monkey-keeper. To him it was perfectly clear that the main objection to the theory of selection was not scientific but emotional, deeply psychological, an ape-complex affecting European man which still exists in large measure today.

It is precisely because apes, and particularly the great apes, have such high intelligence, such a wide range of emotions and are often so astoundingly human in appearance and behaviour, that they are taken by some people to be

Human baby on its belly

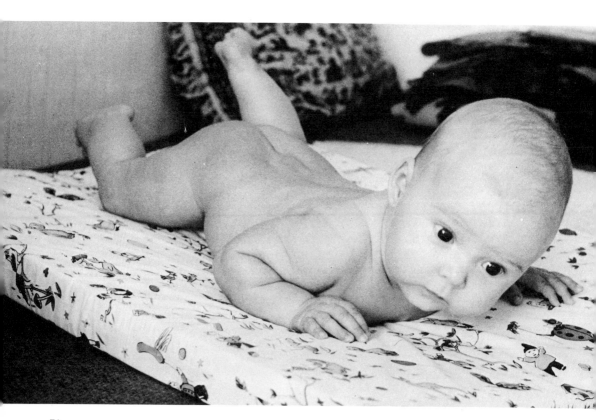

almost embarrassing caricatures of themselves. Depth psychologists say that the ape-complex is derived from this, that man sees himself reflected in the ape, indeed often very sharply reflected, the mirror revealing those habits, characteristics and behaviour patterns which we do not like to admit in ourselves and which we avoid having to recognize by setting up a barrier of taboos.

Many eminent naturalists are themselves inhibited by this ape-complex. Amongst them was no less a man than Lorenz Oken, the brilliant though eccentric biologist of Goethe's time, who had some notion of evolution even then. Yet Oken could not abide apes:

'Apes resemble man in only their bad habits and behaviour. They are wicked and deceitful, spiteful and indecent, adept at learning tricks, but disobedient, often breaking up

Gorilla baby on its belly

Zoo attendants receive a visit from their dear relations

a game by foolery in the middle of it, like a clumsy clown. There is not one virtue which man can admit the apes to possess, still less a use they could have for man. . . . They are merely the bad side of mankind, from the physical as well as the moral viewpoint.'

Today every serious naturalist smiles at this moral rebuke Oken delivered to his animal cousins. The American anthropologist, Ernest Albert Hooton, described the ape-complex with bitter irony in his book *Up from the Ape*, published in 1946: 'I do not see how nowadays any man can look a monkey in the face and claim any sort of relationship to it on the ground of his behaviour. Any decent monkey would repudiate the suggestion of a common ancestral kinship with mankind.'

A second argument against Darwin's theory was more profound: no fossil remains had yet been discovered which could prove the link between apes and man. Darwin did not consider this very important, and was content to bear in mind that 'those regions which are most likely to afford remains connecting man with some extinct ape-like creature have not as yet been searched by geologists'. Vogt took up the issue more thoroughly and pointed out that prehistoric finds had in fact already been made, even in Darwin's time: 'Indeed remains of extinct animals intermixed with human bones had already turned up here and there; but these had either been pushed aside or entirely ignored, or were explained in a manner which hardly cast a favourable light on the sagacity of the observer.'

But Vogt had to admit that even he was not in a position to produce the missing link. 'But when it is added that no intermediate forms can be found,' he writes, 'the history of the last ten years, with all its discoveries relating to man and ape, tells a different tale. Twenty years ago fossil apes were unknown, and now we know of nearly a dozen; who can tell that in a few years time we may not know fifty? . . . Who can say that in ten, twenty or fifty years we may not possess a whole series of intermediate forms between man and ape?'

The search for the missing link began to obsess all
naturalists concerned with the history of human origin. It is
true that not only the Neandertal man but other fossilized
remains had been found well within Darwin's time. But
these finds could scarcely be expected to arouse the interest
of the many contemporaries of Darwin who either believed
that man had existed for no more than a few thousand years,
or found the idea of a close relationship with apes embarrassing. Even such an expert as Sir John Evans, President of
the Anthropological Institute of Great Britain, wrote in
1877:

Ernst Haeckel, reformer of zoology

'There is little doubt that one day one of these early members of the human race will eventually be found. But in the meantime each successive discovery must be received in a cautious though candid spirit, even if eventually we have to carry it to what is called in the City a "suspense account";

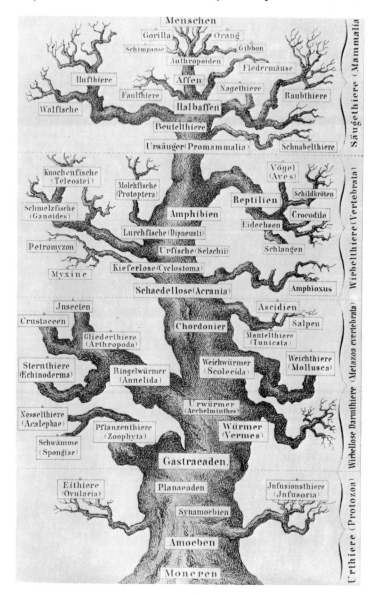

but looking to the many sources of doubt and error which attach to isolated discoveries, I cannot but think our watchword for the future must be caution, caution, caution!'

Nevertheless, long before this time, a naturalist of even greater personality than Karl Vogt had gone so far as to draw up the first hypothetical genealogical tree of men and apes. This was the German biologist, Ernst Haeckel, a 'reforming zoologist', as his contemporaries called him. The fundamentals of his genealogy still hold good today. Haeckel, professor of zoology at Jena, was another ardent defender of Darwin's doctrine and the sarcasm with which he ridiculed the ape-complex far exceeded that of Vogt. 'It is a most interesting and instructive fact', he wrote, 'that it is precisely those people whose intellectual advancement is the least far from that of apes who are the most indignant about the discovery of the natural evolution of the human race from the genuine apes.'

But Haeckel did more than draw up genealogical trees. In a truly prophetic fashion he demonstrated to his specialist colleagues in *The History of Creation*, probably his most important work, that the most likely place of origin of the creatures transitional between ape and man lay in the Asiatic or African tropics of the Tertiary era. The way in which this prophecy was to be fulfilled can be seen in the third and fourth sections of this book. When Darwin read *The History of Creation* he admitted modestly that: 'If this work had appeared before my essay had been written, I should probably never have completed it. Almost all the conclusions at which I have arrived I find confirmed by this naturalist, whose knowledge on many points is much fuller than mine.'

The ancestors of man should not be sought in the Ice Age but much further back, Haeckel had maintained as early as 1866 in his *General Morphology*. They would be found in the Tertiary strata, that is, those strata which as far as our knowledge goes must be several million years old. After Haeckel had read Darwin's work he too, like Huxley, made that momentous pronouncement which in lay circles is often

Genealogical tree of the animal world from single-cell creatures to man, from Ernst Haeckel's 'Anthropogenie oder Entwicklungsgeschichte des Menschen', 1874

laughed at or contested even today: 'It is now an indubitable fact that man is descended from apes.'

Haeckel called this a 'special deductive conclusion' derived from the general inductive law of the theory that 'man has developed gradually out of the Lower Vertebrata, and more immediately out of Ape-like mammals'. Indeed, at the start, Haeckel knew of no palaeontological discovery which could fill the gap between apes and the sophisticated men of the Ice Age, so he included a 'speechless ape-man', *Pithecanthropus alalus*, in his evolutionary tree. This species of man would have lived in the Pliocene, he thought; that is, the final part of the Tertiary era. He had evolved out of a branch of apes and from him arose speaking man.

Haeckel's 'speechless ape-man' was merely a working hypothesis, the nature and use of which had become established long before Haeckel's time as an entirely legitimate method of scientific hypothesis. Moreover Haeckel, despite his usual aggressiveness, was just as cautious as Darwin in his choice of words when approaching the tricky question of man's origin:

'I must here also point out, what in fact is self-evident, that not one of all the still living apes, and consequently not one of the so-called man-like apes, can be the progenitor of the human race. This opinion, in fact, has never been maintained by thoughtful adherents of the theory of descent, but has been falsely attributed to them by their unthinking opponents. The ape-like progenitors of the human race are long since extinct. We may possibly find their fossil bones in the Tertiary rocks of Southern Asia or Africa.'

Later events were to prove this prophecy astoundingly accurate. It is interesting that even Haeckel's opponent, Rudolf Virchow, couched the thesis in similar terms: 'Large areas of the earth have not yet revealed their fossil treasures. Amongst these are precisely the habitats of the anthropoid apes: tropical Africa, Borneo and neighbouring islands, which are still totally unexplored. A single discovery may change the perspective of the whole question [of the origin of man].'

Opposite: Embryo of pig, cow, rabbit and man at three successive stages in their development, from Ernst Haeckel's 'Anthropogenie'

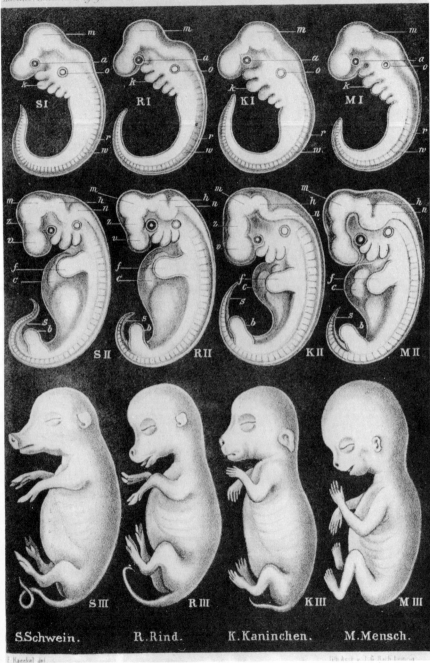

Despite this quite considerable agreement between the opposing scientific views, Virchow made much sport of Haeckel's 'speechless ape-man'. To endow a creature which no one had yet proved to have existed with a zoological name seemed to him a mockery of science. But Haeckel was not put out by this. He even instructed the distinguished portrait painter, Gabriel Max, to paint some pictures of the connecting link as Haeckel imagined him to have looked.

'It is a very remarkable picture', the later explorer and interpreter of Asiatic prehistoric man, Gustav Heinrich Ralph von Koenigswald says of one of these paintings. They are all now in the Haeckel Museum in Jena. 'Under a tree a woman with long lank hair sits cross-legged suckling a child. Her nose is flat, her lips thick, her feet large, with the big toe set considerably lower than the rest. Beside her stands her husband, fat-bellied and low-browed, his back thickly covered with hair. He looks at the spectator good-

Ernst Haeckel comparing a human skull with the skeleton of a gibbon, which he regarded as the ape most closely related to man

*The 'speechless ape-man', Pithecan-
thropus alalus, as conceived by Ernst
Haeckel, and drawn by Gabriel
Max*

naturedly and unintelligently, with the suspicious expression
of an inveterate toper. It must have been a very happy
marriage; his wife could not contradict him, for neither of
them could speak.'

The way Haeckel's 'invented' *Pithecanthropus* was later
actually found is told in the third section of this book. But,
to begin with, interest was focused less on hypothetical
'ape-men' than on the phenomenon of the Ice Age, and with
it the discovery of Ice Age man and Ice Age art.

Part Two

Artists of the Ice Age

The question of all questions for mankind – the
problem which underlies all others and which is
more deeply interesting than any other – is the
ascertainment of the place which Man occupies in
nature and of his relations to the universe of
things.
THOMAS HENRY HUXLEY

LARTET AND CHRISTY

MORTILLET

MARCELINO DE SAUTUOLA

CARTAILHAC

BREUIL

OBERMAIER

KÜHN

CABRÉ AGUILO

HERNANDEZ PACHECO

KLAATSCH

HAUSER

SCHWALBE

BERKHEMER

MARSTON

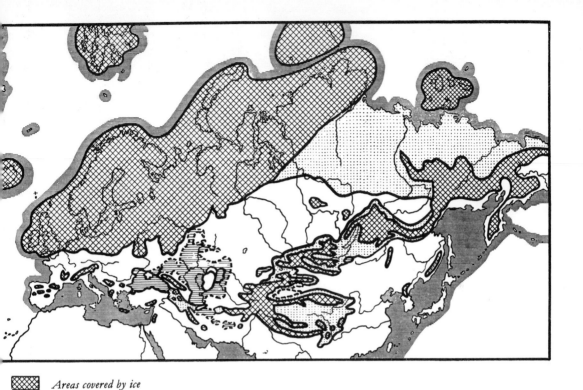

	Areas covered by ice
	Tundra and cold steppes
	Lakes and salt steppes

Europe and Asia during the Ice Age

'Anyone who has had occasion to publish some work on the Ice Age finds to his horror that innumerable manuscripts in terrifyingly bulky parcels accumulate on his desk, with or without return postage, whose senders invariably proclaim that they too have found a solution to the Ice Age.'

This groan was uttered by the German naturalist Wilhelm Bölsche around 1920, showing that even then not much was known about the phenomenon which William Buckland had identified with the biblical deluge and which we refer to today as the Ice Age. There have been many ice ages in the history of the earth, but it is only the epoch immediately preceding our own presence on earth which has any relevance for the history of man's development. In the old days they called it the Diluvium, the time of the flood; now we call it the Pleistocene. In fact, it consists broadly of four glacial periods with three warm periods in between. Formerly it was considered to have begun 600,000 years ago and to have ended about 20,000 to 12,000 years ago – depending on the region in question. But recent research leads us to assume that it began very much earlier. The findings of the American researchers David B. Ericson and Goesta Wollin suggest that the cold period may have set in as far back as two to three million years.

Closely associated with the study, as yet uncompleted, of the Ice Age is the search for Ice Age man. The first prehistorians to know of the existence of the Ice Age and an Ice Age culture had no idea that the Pleistocene comprised four glacial periods, nor of the great duration of each of these periods. Edouard Lartet, a French lawyer, was exploring prehistoric caves and had discovered the remains of prehistoric man well before Darwin ventured to make public his theory. In 1856, in Saint-Gaudens on the north face of the Pyrenees, he made a discovery which attracted little attention then but which was in fact of tremendous scientific importance: he dug up a humerus and a few fragments of skull of a surprisingly human-like ape which he called *Dryopithecus* and which we now know to have lived in the Middle Tertiary. Not until Darwin and Haeckel came upon the scene was *Dryopithecus* allowed to come into its own; indeed for some time it was even regarded as the much sought-after missing link.

However, Lartet's main area of activity was another region of France. A truffle-hunter had told him of the strange remains of bones in the caves and grottoes of the French province of the Dordogne, and he joined up with Henry Christy, a supporter of Lyell, who financed their excavations. These two men, a wealthy English banker and a French lawyer, both inspired laymen like Boucher de Perthes, opened the way to a new undreamed-of world. Indeed it was not mere luck but a sound instinct which made them begin their excavations near the small town of Les Eyzies in the Vézère valley, which has since become the focal point of European prehistory.

The major periods of glaciation during the Pleistocene epoch

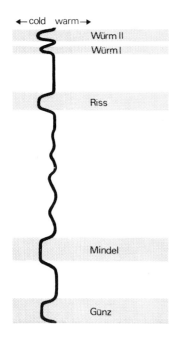

←cold   warm→

Würm II

Würm I

Riss

Mindel

Günz

Engraved mammoth tusk

Their attention was soon attracted by the numerous ledges protected by overhanging rock which characterized this valley. These rock-shelters, known as *abris* in French, provided places in which Stone Age man could have lived in relative comfort. In many of them the debris left by their early occupants was found to have acquired a depth of around twelve feet. Not only flint implements were found there, but ash, charcoal, animal bones, and other things besides, all indicating human activity and the use of fire. The bones were those of mammoths, woolly rhinoceroses, cave bears, reindeer and other Ice Age creatures. Thus man must have lived alongside these beasts.

The cave of La Madeleine even yielded an object which Ice Age man had actually fashioned himself. This was a carved mammoth tusk, one of many relics of prehistoric art which Lartet and Christy brought to light during their excavations.

Engravings of animals on bone and ivory found by Lartet and Christy between 1865 and 1875

At the Paris Exhibition of the summer of 1867, two years after Christy's death, Lartet exhibited the numerous finds he had made in the Vézère valley: stone implements, carved animal bones, in particular such evidence as he had found of early man's artistic activity. It was there that anthropologists and archaeologists first coined the term 'prehistoric'. Now there was no longer any doubt that man's prehistory extended right back into the Ice Age. Moreover, Lartet's evidence revealed that the men of that time were not just primitive barbarians, but beings possessed of intelligence and skills.

But what had they looked like, these artists of the Ice Age? In 1860 Lartet had come across some human remains in a cave at Aurignac; but the experts would not accept his interpretation of them as the bones of Stone Age man. Two years before he died, however, he had the satisfaction of witnessing his son Louis discover some skeletal remains which nobody could deny were those of late Ice Age men, accompanied by some animal bones and flint implements.

The first prehistoric murder: such is the name sometimes given, if rather sensationally, to the mystery of Cro-Magnon. At the back of a rock shelter, now incorporated in the Cro-Magnon hotel, Louis Lartet dug up the skeletons of four men, a woman, and a not quite full-term baby. Obviously these had been the victims of some prehistoric tragedy, for they had been murdered. The female skull revealed a wound which had undoubtedly been made by some cutting instrument; and the other skulls showed signs that violent blows had been dealt them too.

Neither Lartet nor other prehistorians could solve the mystery of Cro-Magnon. But Louis Lartet pictured the life and doings of these Ice Age men of the Vézère valley as follows: 'There is evidence that numerous stays were made in the Cro-Magnon caves, all by the same hunting folk who will perhaps have first used the cave as a shelter when they

Skull of a Cro-Magnon man, front and side view

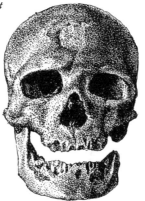
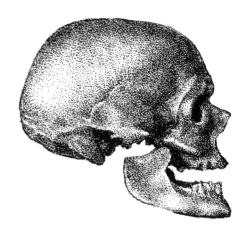

were out hunting, meeting together there to share out the spoils of the chase. Later they came to live there. Then, finally, as their collection of refuse mounted up, gradually raising the level of the floor till it was four feet higher than it had originally been, the living area was so diminished that the cave was no longer comfortable to live in. So they gradually moved away, returning only that one time to dispose of their dead. Since then the cave has no longer been accessible, except perhaps to foxes. The progressive weathering of the surface had provided the remarkable grave with a thick covering, which itself is evidence of its great age.'

Nevertheless Cro-Magnon man was a disappointment, at any rate to those scientists who had taken Darwin's theory too literally and who expected the men of the late Ice Age to have been ape-like early human beings, if human beings at all. For at the time of the discovery it was not generally recognized that man had taken so long to develop. Also, no method had yet been invented by which an accurate dating of palaeontological and prehistoric finds could be made.

Darwin's opponents triumphed for a while; for the habitat of Cro-Magnon man revealed quite clearly that his had not been an animal-like existence. Quite the reverse: he had been a particularly advanced member of our own species.

Geoffrey Bibby, a prehistorian of our own day, describes a typical representative of late Ice Age humanity as follows: 'He was not in the least ape-like. On the contrary, he was fully human – only more so. Above average height – the males approached six foot six – he was shown to have had a broad high forehead, prominent cheek-bones, and a pronouncedly firm chin. His skull capacity was above the average for modern Europeans. If he was the ancestor of modern man – the view now accepted – there would appear to have been a process of degeneration from that point to the present day. . . . For many years anthropologists preferred to believe that Cro-Magnon man had died out, leaving no descendants.'

Since Lartet's find, Cro-Magnon men and members of related late Ice Age races have been discovered in great numbers throughout Europe, Asia and Africa. The old theories of man 'degenerating' and Cro-Magnon man leaving no descendants have long been relegated to our outdated files. There is no longer any doubt that all or at least most types of modern man are descended from the Cro-Magnon race.

A many-sided French scholar, Gabriel de Mortillet, who had taken an active part as a Republican in the February revolution of 1848 and who had lived as an emigré in Switzerland into the sixties, now quickly became the leading prehistorian of his country. He divided the Old Stone Age cultures according to the typical tools used in each period, successfully correlating these individual stages of human culture with particular periods in the earth's recent history. Present-day prehistorians have retained this classification, at any rate in its fundamentals, though certain chronological adjustments have been made.

But the greatest surprise which the Cro-Magnon men sprang on modern science was the discovery of their cave pictures, although the views currently dominating anthropology at first prevented people from grasping its full significance. However, this discovery was a real sensation, and believed by many to be yet further proof of the falsity of Darwin's doctrine. The Spanish prehistorian, Jesús Cárballo, described the feelings of those scientists who had gradually come to accept the contemporary interpretation of Darwin's theory by which early men could have been no more than apes, and who were now forced to come to terms with the existence of a highly developed prehistoric art: 'Primitive man was little more than a gorilla, incapable of any conception of art or science. Besides, how can one make it sound credible that pictures thousands of years old, painted in powdered ochre, could have survived all this time in a deep, damp and pitch-dark subterranean cave?'

A Stone Age cave painter and his work

93

YEARS	ERA	PERIOD DURATION	SOIL CLIMATE	ANIMA
10,000	Holocene	Recent past	Later alluvial soils	Contempo fauna
20,000		Post-glacial	Steppe climate	Steppe ani
100,000	PLEISTOCENE (OR DILUVIUM)	Fourth or Würm Ice Age (97,000 years)	Formation of later loess, cold, occasionally temperate	Reindee Elk Bison Musk-o Wild hor Mammo Woolly rhino Cave-be
		Riss-Würm Interglacial (65,000 years)	Large-scale erosion of soil Multiplication of caves, warm	Cave-lio Cave-hye Elephar Rhinocer
200,000		Third or Riss Ice Age (53,000 years)	Lower layers of caves formed, earlier loess, cold	Mammo Woolly rhino Cave-be
300,000 400,000		Mindel-Riss Interglacial (193,000 years)	Large-scale erosion and removal of soil terrace formation temperate to warm	Antique ele Mercks rhino Primitiv hippopota Sabre-tooth
		Second or Mindel Ice Age (47,000 years)	Cold	Mammo Woolly rhino
500,000		Günz-Mindel Interglacial (65,000 years)	Alluvial soil on plains, warm	Southern ele Etrusca rhinocer Early wild h Deninger b
600,000		First or Günz Ice Age (65,000 years)	Cold	?
	PLIOCENE	Last Tertiary Period	Tropical temperate	Mastodo Dinotheri Okapi Monkey

Table of the Pleistocene cultures, as drawn up in Henri Breuil's time. Today it is thought that the Pleistocene lasted two to three million years

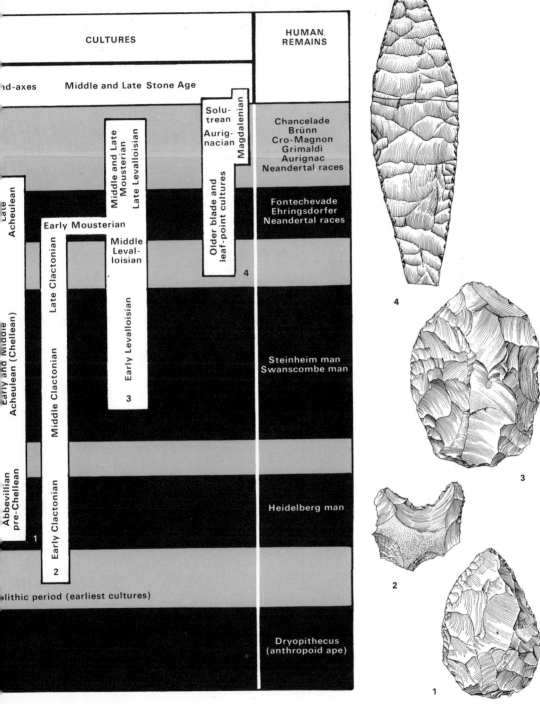

CULTURES						HUMAN REMAINS

nd-axes   Middle and Late Stone Age

Solu-trean

Aurig-nacian

Magdalenian

Older blade and leaf-point cultures

Chancelade
Brünn
Cro-Magnon
Grimaldi
Aurignac
Neandertal races

Fontechevade
Ehringsdorfer
Neandertal races

Late Acheulean

Middle and Late Mousterian
Late Levalloisian

Early Mousterian

Middle Leval-loisian

4

Early and Middle Acheulean (Chellean)

Late Clactonian

Early Levalloisian

3

Middle Clactonian

Steinheim man
Swanscombe man

Abbevillian pre-Chellean

1

Early Clactonian

2

Heidelberg man

lithic period (earliest cultures)

Dryopithecus
(anthropoid ape)

4

3

2

1

95

It all began with a dog back in 1858; this dog was chasing a fox when it suddenly fell into a crevice and could not extricate itself. The huntsman forced his way into the cleft, saved the animal, and in the process discovered a very large cave, partly blocked by a mass of stone. The cave, in the vicinity of the castle of Santillana sur Mar, near Torrelavega in northern Spain, was situated beneath a hilly meadow on the property of the nobleman Don Marcelino de Sautuola. Today the site is known throughout the world as Altamira.

Contemporary portraits of this Don Marcelino, to whom we owe perhaps the greatest discovery in the history of our time, do not make him look like a scholar at all but like a typical member of the ancient Spanish nobility. Geoffrey Bibby describes one of these pictures: 'A slim, athletic figure, his hair receding from his broad forehead, and the bushy side-whiskers and heavy moustache gave an exaggerated air of authority, emphasizing the firm chin and the level brows, and almost hiding the humorous mouth.' But his appearance was deceptive; for Don Marcelino was an expert on the geology of his own land and was particularly interested in the preservation of outstanding works of nature. Yet it was not until 1875 that he first entered the cave of Altamira with some workmen, had the entrance enlarged, some blocks of stone cleared away, and found inside a number of animal bones which he handed for closer inspection to his friend and professor in geology, Juan Vilanova. Professor Vilanova identified them as the bones of bisons, wild horses and other Ice Age animals; and what was more, men had obviously split the bones to extract the marrow.

For a further four years Don Marcelino did not enter the cave. This was not for any lack of interest; on the contrary, he spent these years well, visiting and exploring other Spanish caves and gathering experience in excavating and salvaging their contents. As the Altamira cave was on his land no one could enter it without his permission so he could safely leave it awhile. He went to the Paris Exhibition of 1878 and studied the drawings scratched on bones and the

Opposite: Altamira initiated the study of cave art. These are bison, executed in several colours on the roof of the cave

*Cave painting in black of a woolly
rhinoceros, from Rouffignac*

stone and bone implements Lartet and Christy had found in the Périgord region of France. He then began a systematic exploration of the Altamira cave, finding along with many more animal bones a number of stone implements comparable with the ones he had seen at the exhibition. The blackened earth, with fragments of bones and mussel shells scattered here and there betrayed that prehistoric man had camped around a fire and sat down there to his meals. There was another, stranger piece of evidence to which he did not at first pay attention – the encrusted deposits of black and red colouring agents found in the mussel shells and among the refuse.

But María's eyes were sharper. One day Don Marcelino took his five-year-old daughter with him into the cave. The child lost interest in her father's doings and played about in the recesses and passageways illuminated by the light of a candle. Many years later, when she was grown up, María de Sautuola described to the Spanish mayor and prehistorian Alcalde del Río what she had seen and how her father had reacted: 'While I was running about in the cave, playing here and there, I suddenly made out forms and figures on the roof. I pointed the pictures out to my father, but he just laughed. Soon, however, he became more interested, held his lamp aloft and saw the pictures all about him. He touched the colour with his finger and found it was real oil paint. He was so excited he could hardly speak. Then he remembered that years ago he had noticed some black markings on the walls but had never thought them important.'

The impression made upon a specialist of today by the cave paintings of Altamira is vividly conveyed by these words of the German prehistorian, Herbert Kühn: 'The bumps on the walls take on movement and life. One of the lumps carries the figure of a standing bison, another swelling bears a beast crumpled up. The prehistoric artist has utilized the irregularities of the surface to convey the impression of relief. Thus the pictures are paintings and relief work at one and the same time, and as such are impossible to render well on paper. Here, for instance, is a

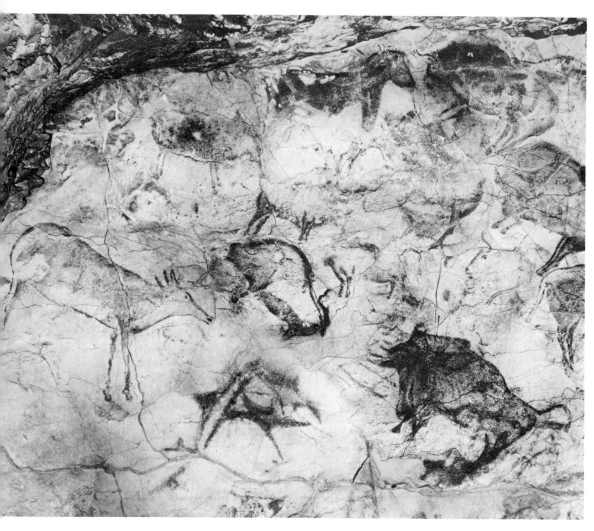

Ceiling paintings in the Altamira cave. A group of bison, around them a hind, a boar, two horses, and another boar

hind. It measures about four and a half feet long, is masterly in execution. And over there is a bison as if struck by a spear, a remarkable picture, quite unforgettable: it is marvellous to see how the beast sinks its head, its hind legs double up, and its coat shines like that of a living animal, and how the figure stands out and the colours have been applied despite all the difficulties presented by the surface. . . . The Ice Age men are very real to us here, very familiar. . . . We are used to thinking of the beginnings of art as stiff and formal – but here we are confronted with paintings that are absolutely natural, which breathe, live and speak.'

Many visitors who have a considerable knowledge of art reckon the Altamira paintings to be among the greatest and most unforgettable works of art they have ever seen. This is not merely because of their great antiquity, but because of their stylistic invention and their very special mastery of form. Don Marcelino must have felt the same; for on the very evening of his discovery he visited his friend Vilanova and together they thoroughly explored the subterranean caves and studied their strange pictures.

At Vilanova's suggestion Don Marcelino wrote a full report of his findings, and had it extensively illustrated. But he made one grave error. An impoverished French painter was living with him at the time, and Don Marcelino commissioned him to do the drawings for the report. When it was published in 1880 none but the curious showed any interest in it. Thousands of people visited the cave; even the King of Spain, Alfonso XII, came to Altamira and inscribed his name in soot on the wall near the entrance. But scholars, for the most part, held back. Only a single expert, the engineer and prehistorian Edouard Harlé, thought Altamira worth a visit. But he did not come to study the pictures. And when he heard that a roving artist had been living at Don Marcelino's castle at the time of the discovery but had since left, he hurried back to France and proclaimed the alleged Ice Age paintings to be nothing but a fraud. Some of his colleagues were more cautious, however, not wishing to accuse so distinguished a professor of geology of being an impostor. They maintained Don Marcelino and Vilanova had in all good faith been taken in by some daubings on the walls done by local peasant boys.

At the Congress of Prehistorians held at Lisbon in 1880, the paintings were pronounced as inauthentic. Vilanova did his best to protect his friend's name, but he received absolutely no support. The entire fraternity of prominent anthropologists, archaeologists and prehistorians were against him. Amongst these were such

Henri Breuil devoted much time to the study of cave paintings. This drawing of his reproduces the animals on the ceiling at Altamira

distinguished men as Rudolf Virchow, the Swedish archaeologist Oscar Montelius, the English Darwinist John Lubbock, and the eminent French scholar, Émile Cartailhac. Cartailhac's condemnation was very sharp, and most participants supported him. This Ice Age art was nothing more than a 'caricature, a swindle, a joke', he declared, cooked up to dupe gullible scientists. Even Lubbock did not believe that the pictures could be genuine, for he could not free himself of the idea that men living as early as that must have led a semi-animal existence. Vilanova finally capitulated and refrained from mentioning Altamira again.

The master of Santillana sur Mar did all he could to bring the matter up again at a scientific gathering, but he was cold-shouldered. Virchow made sure that his request to speak at the congress of prehistorians at Berlin in 1883 was not granted. Five years later he died; and his former supporter Vilanova outlived him no more than a further five. The only person who believed in the authenticity of these subterranean paintings at Altamira was their actual discoverer, María de Sautuola.

Many caves with paintings could only be reached by negotiating subterranean waters. This photograph was taken at the entrance to the Tuc d'Audoubert cave on 12 October, 1912, two days after the discovery of the bison sculptures. It shows (from left to right) Count Bégouen with two of his sons, Henri Breuil, a third Bégonen son, and Émile Cartailhac

Even in Don Marcelino's time, however, caves were found in France whose walls were decorated in much the same way as the Altamira cave. On 6 May, 1881, near the village of Marcamp in the province of the Gironde, some peasants came across a hole in the earth which they first assumed to be a badger's sett. They widened the entrance, and soon found themselves in a great hall-like cavern. Two and a half years later the prehistorian François Daleau made a thorough investigation of the cave, now known by the name Pair-non-Pair, and noticed some marks scratched on the walls. Unfortunately he did not look at them closely. Not until 1896, after many other cave paintings had been discovered in France, did he remember the decorations on the walls. He had the cave completely cleared, and found twelve pictures of wild horses and bison upon its walls, still bearing traces of colour. Even so, Daleau had been so strongly influenced by the violent attacks upon the authenticity of such paintings that he did not dare mention his find.

Engraving of reindeer and fish on a reindeer antler

The next discovery was also kept dark. Between 1881 and 1884 a clergyman with a great enthusiasm for the study of early history completely excavated the Marsoulas cave, situated near Salies-du-Salat in the province of Haute-Garonne. He was constantly coming upon pictures on the walls, but, believing the experts, took them to be recent and made no mention of them in his report of the cave.

These men who first discovered prehistoric paintings cannot be blamed for their scepticism and silence, for certain unfortunate happenings had brought the whole subject into disrepute. These included the first forgeries of prehistoric art. In 1874 a schoolteacher, M. Merk of Basle, had found the splendid engraving of a grazing reindeer in the Kesslerloch cave near Thayngen on the Swiss-German border. We now know this drawing is unquestionably authentic. But, as over the next few years he explored deeper and deeper into the cave, Merk came across many more similar drawings, among them those of a bear and of a fox. When Merk published illustrations of these, the director of the Mainz

A standing bison, incised on stone

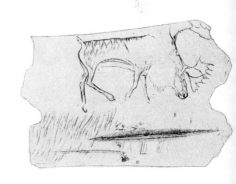

Engraving of reindeer on reindeer antler, found in the Kesslerloch, near Thayingen

museum immediately recognized them as copies taken from an illustrated children's magazine which had printed pictures of a large number of zoo animals. Seeing how delighted the Swiss teacher had been to find the original engraving, the men working for him had felt impelled to provide him with more exciting finds: they had commissioned a schoolboy from Schaffhausen to scratch copies of these children's magazine pictures on the walls of the cave.

After this unfortunate incident, even those who still retained some belief in the authenticity of at least some of the wall paintings discounted the idea that they came from the distant prehistoric past, believing them to have originated in classical antiquity under Greek or Roman influence. Distinguished prehistorians declared that sooner or later someone would find paintings accompanied by Greek or Latin letters, and matters were left at that.

In the valley of the Dordogne, that paradise of prehistoric man, where Lartet and Christy had once done so much excavating, the local inhabitants for their part were happily unaware of such ideas and prejudices. A certain M. Berniche had made himself a storage room, partly walled, out of a cavern called La Mouthe, near Les Eyzies, the back of which extended through a narrow crevice into a passage at first deemed impassable. But on 11 April, 1895, a schoolboy, Gaston Berthoumeyrou, squeezed through this hole and discovered a number of strange engravings on the walls beyond. He told his teacher of the discovery, and the news finally reached the well-known archaeologist, Emile Rivière. Rivière systematically explored the La Mouthe cave and noticed that some parts of the pictures disappeared beneath layers of chalk and stalagmite. Since such deposits could only have been formed over a period of thousands of years, the great age of the paintings was proved.

Rivière was convinced that if he were to make public his belief that these paintings were old he would immediately incur the bitter hostility of the great Cartailhac, who had meanwhile been appointed president of the Prehistoric Society of France. He therefore by-passed the society and

Part of a painting in the La Mouthe cave, reproduced in line

went straight to the Paris Académie des Sciences, where they were rather more open-minded than the immediate specialists in the subject. Its members at least listened amiably to Rivière's arguments in support of the paintings' great age. But it seemed that in the meanwhile Cartailhac himself had become less sure of the correctness of his original verdict. A note about the Teyjat cave, which had been discovered in 1889, was found among his papers after his death: 'I get the impression that there could have been drawings and paintings on the wall of the cave after all. Will speak to Breuil about it.'

At that time Henri Breuil was in his early twenties – a young Abbé with a tremendous enthusiasm for prehistoric finds, the precursor of many such in France and Spain. On Cartailhac's recommendation he had taken part in the investigation of a number of French caves, and it had clearly occurred to him before it did to his instructor that these controversial paintings could be genuine after all, and very ancient into the bargain. It was an auspicious moment in the study of prehistoric art when in 1901 Breuil penetrated the cave of Les Combarelles, situated near La Mouthe.

Les Combarelles was on the property of the same M. Berniche who had converted La Mouthe into a storage place for his farm; and as chance would have it the same Gaston Berthoumeyrou who had discovered the pictures of La Mouthe had explored the passages of Les Combarelles long before Breuil's visit. Even Emile Rivière had done some excavating there, and had found an engraved reindeer bone. But neither Berthoumeyrou nor Rivière had noticed pictures in the cave; these had been seen only by its owner, M. Berniche.

On 9 September, 1901, Breuil began a systematic exploration of the cave, accompanied by the prehistorians Capitan and Peyrony. Breuil was later often to describe the shouts of joy which went up as they discovered animal after animal on the walls of the cave, true to life, clear, and accurately drawn. They traced the outlines with their fingers; and Breuil immediately set to work copying the pictures so that

Horse engraved on the wall of Les Combarelles cave

he could pass on his immediate impression to the outside world.

Shortly after this Peyrony, who was teaching in Les Eyzies at the time, found more large coloured paintings on the walls of the neighbouring cave of Font-de-Gaume. Breuil and Capitan made a study of these as well. When the Académie des Sciences published its first report of these discoveries, Cartailhac's opposition broke under the weight of the accumulated evidence. The profound emotion he felt as a result of his errors can be seen from the title of the recantation he immediately published: *Mea culpa, d'un sceptique!* The former sceptic not only acknowledged his guilt but did all he could to make amends for the injustice done at Altamira.

'We are living in another world!' Breuil and Cartailhac are reported to have exclaimed on their first visit to Altamira. Quite contrary to the attitude of so many scientific dogmatists who are unwilling to admit their errors, Cartailhac visited the daughter of the man who had discovered the paintings and made her a formal apology. Only death could relieve him of the burden of his guilt, he is reported to have exclaimed. María de Sautuola smiled. She had a very special surprise for her visitors. Spanish amateur explorers, among them Alcalde del Río, had discovered a large number of cave paintings over the preceding years, and although they had written reports about them they had discreetly kept these to themselves. When Cartailhac asked why they had let nothing be known of their discoveries, María de Sautuola shrugged her shoulders: 'But you wouldn't have anything to do with cave paintings!' she exclaimed.

The discovery of Font-de-Gaume has been justly described as one of the greatest events in the remarkable story of our uncovering of prehistoric art. It gave the Abbé Breuil the incentive to devote his life to the study of this earliest manifestation of human art, and he became the acknowledged expert on the subject. By this time Cartailhac too was singing the praises of our Ice Age ancestors' artistic talents: 'These undreamed-of paintings of the Quaternary era are tremendous in every respect, and surpass all ethnographic parallels. However distant these ancient people might be from us in time, we are drawn to them and feel closely related by sharing the same devotion to art and beauty. We should not blush to call them our ancestors.' Added to this was the astounding artistic expressiveness of

Top: head of a horse, engraving in the Font-de-Gaume cave

Bottom: Two horses' heads carved out of reindeer antler from Mas d'Azil

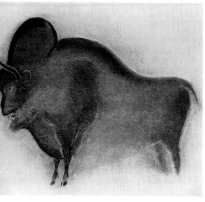

Reproductions of two famous cave paintings from Font-de-Gaume, by Henri Breuil
Top: 'the little mammoth'; below this, a bison

these pictures. Although they had been painted by nomadic hunters who knew neither how to till the soil, breed cattle, nor how to fashion pots or wheels or carts, they can be compared with civilized man's greatest works of art.

Herbert Kühn describes his first impression of the cave of Font-de-Gaume thus: 'The pictures differ very much from one another in dimension. Some are almost life-size, nine feet tall, others much smaller. Often important parts of the body have been gently emphasized by engraving. Many pictures have been executed entirely in black or red, but in others a variety of nuances in colour give the impression of a number of different tones. Unevennesses, dips and protuberances in the stone have often been used to bring out particular parts of the body in relief. When you stand in the midst of it, look all around and take in the chamber in its entirety, you are immediately aware that you are in a sacred place, a cult place, where Ice Age man underwent his religious spiritualization.'

The religious significance of these pictures was and still is disputed. It was natural that the many churchmen who explored these caves so enthusiastically, Henri Breuil and Hugo Obermaier in particular, should take them to be the first evidence of a prehistoric religion, even an indication of an ancient monotheistic belief. The more materialistically inclined prehistorians have often criticized this entirely theological interpretation. Gabriel de Mortillet, the first to classify Ice Age cultures, warned Cartailhac before the latter's visit to Altamira: 'Beware the snares of the Spanish clergy!'

One of the most gifted but because of his methods most controversial discoverers of prehistoric remains was Otto Hauser. He was convinced that Cro-Magnon men had not produced these artistic works of genius but had been in many cases responsible only for the amateurish scribblings around them. He accused Breuil of touching-up the pictures, and even of outright forgery. In his memoirs Hauser describes in his own polemical fashion how he had marked a few particular cave drawings with clay, only to find that

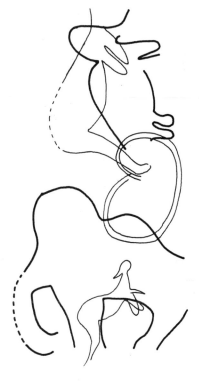

Cave-artists also used fingers to execute their designs. This depiction of a mammoth and a woman in the Pech-Merle cave is an example. Left, a detail of the original; right, outline of the whole group.

after a visit by these 'cave-exploring priests' his lumps of clay had been painted over and the paintings themselves improved. This is clearly malicious gossip, as is also this quarrelsome man's assertion that on many of his visits he had touched the paintings and found fresh colour on his fingers.

There have been constantly recurring rumours, right up to the present day, that various cave paintings or even whole caves of paintings are not authentic, forged partly to undermine academic theories and partly to attract the tourist trade. On how shaky a foundation suspicions of this kind rest can be seen from the case of the famous mammoth pictures of Pech-Merle. This was the occasion when no less

107

In the Pech-Merle cave, alongside his renderings of wild horses, Ice Age Man left us silhouettes of his own hand

a person than the French writer André Breton, one of the spiritual fathers of surrealism, was charged with describing one of these mammoths as a forgery, and the court of the tiny town of Cahors had to decide whether the trunk at least of the mammoth was genuine. Scientists thoroughly studied the caves, and André Breton's opinion was rejected on the basis of their evidence. He was obliged to pay nominal damages.

The Rouffignac cave, covering an unusually large area and extending back more than nine miles, had been visited by all sorts of people for over four hundred years before pictures were discovered there. Tourists had inscribed their names on the walls in soot. Then, in 1956, the Plassard family, who owned the cave, believed they could discern faint drawings beneath the signatures of the visitors who had wandered curiously through it. After considerable trouble the experts managed to clean off most of the modern *graffiti*, revealing innumerable pictures of mammoths, bison, horses and other Ice Age animals, including about ten drawings of the woolly rhinoceros. The Abbé Breuil, his opinion sought, judged them to be genuine. But, because drawings of woolly rhinoceroses had so seldom been found in Ice Age caves, the greater part of the specialist world doubted whether the old French master could be right this time. However, before a scandal could develop, Breuil's verdict was proved, the woolly rhinoceroses themselves providing the evidence. Not only did chemical tests confirm their great age, but also the actual representation of the animals was so accurate that no one, apart from a zoologist of pronounced artistic talent, could have produced such precise renderings of this beast. But why had Ice Age man chosen to draw woolly rhinoceroses? Why did he draw at all?

It was cave paintings which first showed us what the woolly rhinoceros had really looked like. These are from Rouffignac

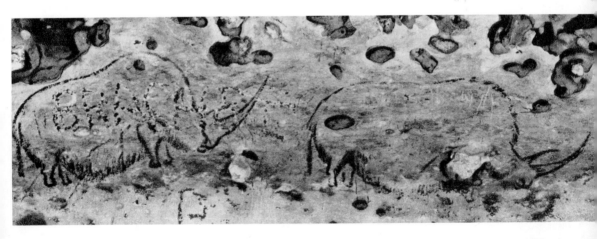

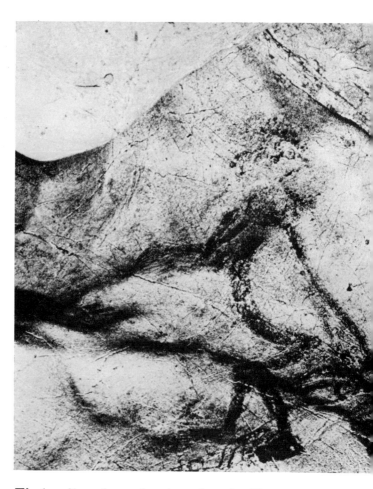

'*The Sorcerer*'; *representation of a man in an animal skin in Les Trois Frères cave.*
Left, drawing by Henri Breuil; right, the figure as traced in black on the wall of the cave

That cult and magic played a significant role in the lives of their creators is evident from many of the paintings to be found in these caves. The pictures found in the halls of the Tuc d'Audoubert cave show quite clearly that early hunting tribes had used this place for initiating their adolescent children into adulthood, a custom still practised by many primitive tribes today. These rites instructed them in the secrets of hunting and the propagation of their own kind. A sculpture of two bison to be found there shows the bull about to mount the cow, and round about the sculpture there is still evidence of the footmarks of adolescent youths who had presumably circled it in performing some ritual dance.

High up in the cavern of the Les Trois Frères cave we find the strange picture of a sorcerer dressed in an animal skin, also indicating that rites of some sort were performed there, probably hunting rites. The manner in which the animals are depicted makes this highly probable: arrows are flying in their direction, their bodies exhibit open wounds, while blood can be seen gushing from a bear's mouth. Breuil's verdict on this cave was as follows: 'One involuntarily thinks of the puberty rites which play such an important part in the life of a tribe, when boys, newly men, become accepted as such.'

Innumerable drawings of a similar nature point to prehistoric initiation, hunting or fertility rites. Beasts are represented as coupling, pregnant, giving birth, or suckling. Women's sexual characteristics are strongly emphasized, and there are strange perplexing pictures of lovers, or of women giving birth, whose outlines follow the outlines of the body of a bear, deer, or other powerful animal. Whether all these very widely varying representations should all be interpreted as having some magico-religious significance, as is customary today, is uncertain. These prehistoric men probably painted out of a variety of motives, not merely out of joy in decoration and beauty, not merely for cultic or religious reasons, but from an urge in which all these motives play their part. Indeed the art of later cultures arose precisely thus.

Just as art historians had devised their own terminology and system of classification for the different art styles and epochs, so prehistorians now followed suit. Today they take much pride in the tremendously subtle distinctions, and distinctions within distinctions, they have evolved in order to lend some chronological order to the cultural and artistic development of Ice Age man. The conception upon which this chronology is based is the same as that used by the first great geologists and palaeontologists; just as epochs in the history of the earth are determined by the characteristic fossils of the period, so individual cultures are determined by the particular kind of stone implements associated with them and the method of their making, and also by the typical characteristics of their art.

This system is based on the original classification drawn up by Gabriel de Mortillet, who gave the various steps and stages of Old Stone Age culture the often tongue-twisting names of the small French towns where the first finds had been made. The earliest periods of Mortillet's system, the Abbevillian, Chellean, Acheulean and Mousterian, are distinguished by the crudeness of their hand-axes which gradually improve in technique as time goes on; but as no pictures or other works of art have been discovered from these periods, which according to modern dating go back 100,000 to 500,000 years, they are of no concern to the art historian. But when we get to the cultural periods of the last glaciation, the Würm Ice Age, we find three quite clearly definable artistic styles which can be directly linked to the corresponding geological periods, stone cultures and human finds.

The oldest pictures, which were often touched-up at later periods, and also quite a number of carvings, sculptures and stone figures, belong to the Aurignacian culture, named after the southern French town of Aurignac. This Ice Age man not only manufactured flint weapons, but also made small flattish oval implements, curved beak-shaped awls and long spatulas out of stone, as well as bone, horn and ivory utensils. The first human remains associated with the Aurignacian show that the members of the Aurignac, or Brünn, race differed somewhat in appearance from the tall, broad-skulled Cro-Magnon men: the skull is longer, the body slimmer and the forehead not so high.

There follows the Solutrean, named after Solutré, near Chalons-sur-Saône. Here laurel-leaf-shaped spearheads were found, also the bones of innumerable wild horses upon which Solutrean man seems to have fed. Some human remains were first classified as being specifically Solutrean, but this was soon discounted, for the people responsible for the Solutrean culture differed neither morphologically nor anatomically from those of preceding or subsequent epochs. Solutrean man's sole claim to progress was his improved

Opposite: The imprints of hands, surrounded by marks of pigment (Pech-Merle), the significance of which is still disputed

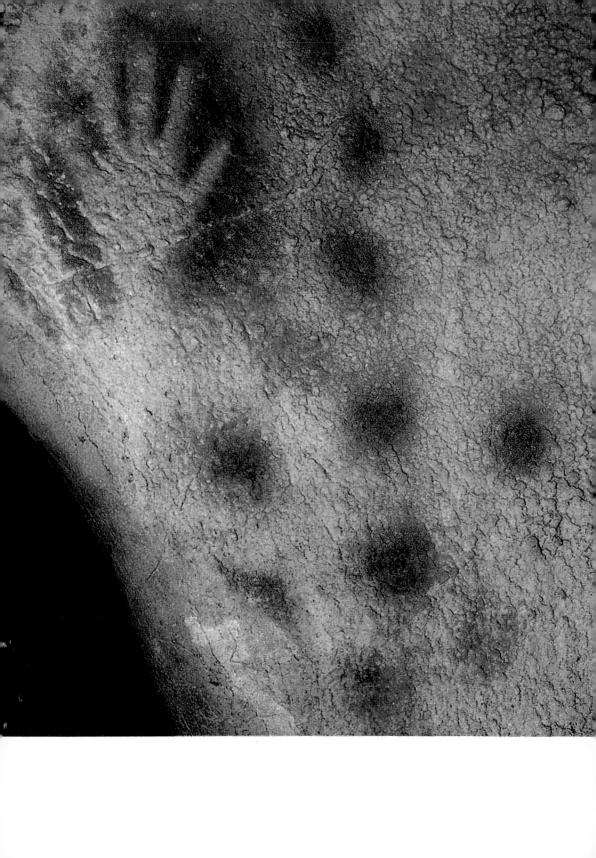

*Representations of bisons in the
main hall at Lascaux*

technique in flint-working; but he made no bone implements, nor does he appear to have possessed any artistic awareness.

The Magdalenian, named after the cave of La Madeleine in the Vézère valley, concludes the Würm Ice Age cultures, and corresponds to the time of Cro-Magnon man. This was the age which produced the finest implements of the Old Stone Age; this was the time, too, when the most artistically advanced cave paintings were executed, or so it was first assumed.

Thus prehistorians adopted the principles fundamental to the theory of evolution, basing their system of classification on the idea of a gradual advance from simple and humble forms to increasingly superior and more perfect creatures, and on a gradual improvement in their manufacturing techniques.

In the Late Ice Age, or Middle Stone Age, the Magdalenian was gradually superseded by the Azilian, typified by the rock pictures we have yet to discuss, by an increasingly abstract treatment of the objects represented in their art, and by the use of bows and arrows, boomerangs, spearthrowers and other sophisticated weapons. The stages which followed on after this period, from the strongly systematized art of the Late Stone Age up to the art of which we have historical knowledge, that of antiquity, fitted easily into this historical picture of evolution.

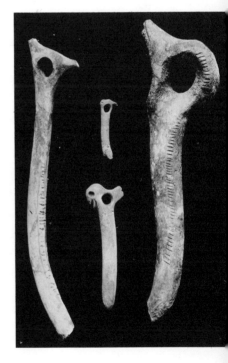

'Bâtons' ornamented with animal heads, from Le Placard (Charente): fox, bird, and ibex

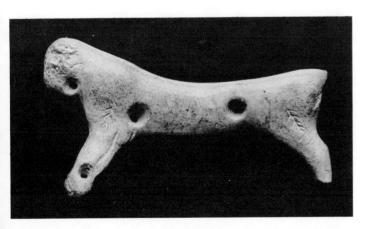

A feline carved out of a reindeer antler, from Isturitz, Pyrenees

In 1940 this system of classification completely collapsed. One warm September day four fifteen-year-old youths were wandering through the woods of Lascaux, on the property of the Comte de Rochefoucauld which lies near Montignac in the Dordogne. Their terrier dog, Robot by name, ran off after a rabbit through a thicket and suddenly disappeared down a crevice in the earth. The boys could no longer see him. One of them, named Ravidat, feared he would get into trouble if they did not find the dog, which belonged to his parents, so they all began a thorough search for it. He described the events which followed in an essay he wrote for his teacher which, despite the schoolboy style of the original, still gives us a good impression of the excitement he felt at the time of the discovery:

'I managed to get five to six yards down into the crevice, head first. Then I switched on my torch and looked around; but I had hardly begun to go forward when I lost my balance and tumbled some distance down. Bruised all over and the breath nearly knocked out of me, I got up and re-lit the torch which had gone out. When I saw that the way down was not too dangerous I called to my companions, telling them to come down too, but very carefully. As soon as we were all together again we began to explore the cave. We crossed a great hall and reached a passage which was narrow but fairly high. And then, as I shone my torch upwards, we saw many traces of colours in the shaky light.

'Fascinated by these shapes, we began to look carefully at the walls and to our great surprise discovered many figures of animals, often very large. Then we realized with a jolt that we had discovered prehistoric paintings. Encouraged by our success we went on through the whole cave, discovering more and more. Our excitement is indescribable. A band of wild men letting rip in a war dance could not have behaved more crazily than we did. Afterwards we made a solemn promise not to tell anyone of our discovery yet, but to return the next day with better torches.'

This, then, was the way the Lascaux cave paintings were discovered – a 'prehistoric Louvre', as they have been called. The most important thing about these pictures is that the oldest of them are not the products of the high refinement of the Magdalenian period, but of a real peak of artistic ability achieved during the much earlier Aurignacian period. They were of just as high a standard and just as impressive as the later, Magdalenian-style achievements. Thus the art historians' calculations were completely upset. They could no longer talk of a progressive advance in artistic achievement during the Ice Age.

Looking into a magic world of Ice Age paintings in the main hall at Lascaux

The renowned animal frieze in the vault of the main hall at Lascaux: a 'unicorn', prehistoric wild horses, aurochs (bull and cow)

The pictures of Lascaux were in a far better state of preservation and their colours far brighter than most other subterranean art so far known. M. Laval, the teacher of these four boys, immediately informed the acknowledged authority on the subject, the Abbé Breuil, as well as other eminent prehistorians. The Academy of Fine Arts in Paris contracted with the owner of the property, Comte Emmanuel de la Rochefoucauld-Montbel, to have the site cleared, steps laid and doors inserted, and the cave opened to the public. One of the first visitors from Germany to describe what he saw was Herbert Kühn. He had produced a full study of Ice Age paintings in 1921, thus raising the subject to the level of a scholarly discipline, no longer the province of amateurs. His report of the Lascaux paintings conveys the effect they had upon an enthusiastic expert who approached them with some artistic knowledge:

'The discovery is one of the most tremendous man has ever made. It is just as important as the discovery of the king's tomb at Aswan, or the opening up of the great temple of Angkor Wat. Indeed, this find touches us at an even more profound level, for it is the oldest art yet found, as well as the most important.'

Prehistorians were now forced to recognize that in the Ice Age too there had been higher and lower levels of artistic achievement which not only succeeded one another but were contemporaneous. Man had not developed in direct sequence, progressing forwards step by step; his development was the result of sharp competition with members of his own species, in the course of which the more powerful cultures and those best adapted to their particular circumstances often overtook their weaker rivals. Prehistorians found that the different types of man existing in Europe in the late Ice Age – the Cro-Magnon race, the Brünn race, the Chancelade race and others too – could no longer be identified with particular stone cultures nor even with particular artistic styles. There was only one difference between the art of the Aurignacian and Magdalenian periods: the older representatives of *Homo sapiens* preferred sculpture and relief work, the later cultures painting and drawing.

The cave-artists of Lascaux depict the dangers of the chase. On the left a woolly rhinoceros, on the right a bison tossing a man to the ground

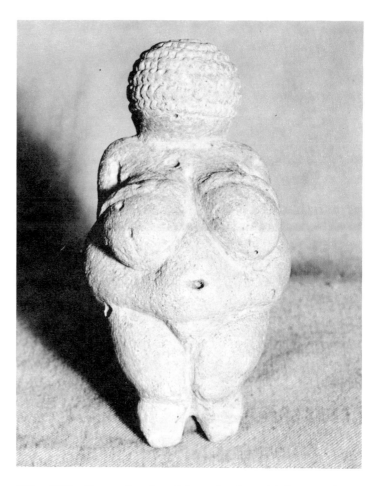

'The little figure is about four inches high carved out of a fine porous limestone, absolutely intact and showing irregular patches of its former red colouring. It is the figure of a very obese woman, with large breasts, a pot belly, broad hips and thighs, but with little in the way of fat on her buttocks. The genitals are strongly marked, the back anatomically correct and furnished with a number of accurate details. Her hair is represented as a series of knots laid in concentric circles around the greater part of her head, but the face has been omitted entirely. There is no indication at all of eyes, nose, mouth, ears or chin. Scant attention has been paid to the upper limbs, the forearms being indicated in shallow relief – mere bands laid across the breasts. The knees are well formed, but the legs are much foreshortened, even though well shaped; and the front part of the feet has been left off. The only indication of drapery or adornment are coarsely knotted bracelets on both forearms.'

This is the prehistorian Szombathy's description of a 'Venus' of the Aurignacian culture which was found by a road workman in 1908 near Willendorf in the Danube valley in Lower Austria. This tiny figure was the first piece of Ice Age sculpture representing the human form to come to light. As with the Ice Age cave paintings, there was much discussion among experts about the purpose and significance of this little Venus.

Those who believed in the cultic and religious nature of Pleistocene art gave full rein to their imagination, seeing in the lowered head and clasped hands those of a woman in prayer. But a less biased verdict was that this Ice Age Venus was far more erotic and sexually potent than her successor, the Greek Venus de Milo.

But it is to the credit of those theologians who were experts in the subject that they concentrated more on her sexual attributes than on the other less strongly represented parts of her anatomy. Breuil's colleague, Hugo Obermaier, immediately declared the figure to be some sort of sexual or fertility idol. 'The figure as a whole reveals that its sculptor had complete artistic mastery of the human form, but was only concerned to emphasize the primary and secondary sexual characteristics of the female. The rest has been brilliantly reduced to an absolute minimum.'

Subsequently other and very different 'Venus' figures of the Aurignacian and later cultures were found, chiefly in the established areas of excavation in France. The Venus of Lespugue is a particularly good example of perfect artistry: it is six inches high, carved in ivory, and the body, apart from its sexual characteristics which are again much stressed, has been brilliantly stylized. Obermaier's pupils, Hans-Georg Bandi and Johannes Maringer describe it as follows:

'Here the rules of composition of the Aurignacian sculpture of women are shown at their most advanced, the figure

Two views of a stylized 'Venus' found near Lespugue

119

nevertheless revealing a really classical balance of proportion. The axial lines, for instance, are uncommonly clear. In the longitudinal axis the body is neatly divided into two exactly corresponding halves through the middle line between breasts and legs; horizontally, too, the upper part rhythmically echoes the lower part, the shoulders corresponding to the powerful thighs, the head to the rounded shanks. In outline the figure is definitely a unit, the arms,

Left: female clay figure found at Dolni Vestonice

Right: fragments of male figure carved out of a mammoth tusk, from Brno

like those of the Willendorf Venus, following the line of the shoulders, with forearms laid across the breasts. Another characteristic especially marked in the Lespugue figure is the emphasis on the central region of the woman's body, by comparison with which all other parts of her anatomy recede into unimportance, intentionally disregarded or intentionally altered in proportion.'

But the Aurignacians did not always ignore the face, as can be seen from the head of a woman found at Brassempouy. Despite its smallness – it is only 1½ inches high – it is the most beautiful portrait of an Ice Age human being that we have. It is a curious fact, for which no explanation has yet been offered, that the Aurignacian period produced such fine sculptures, both realistic and interpretative in style, whereas the Magdalenian period, when wall painting was at its peak, produced statues of such low artistic merit.

Left: ivory head (4·8 cm. high) from Dolni Vestonice
Right: ivory head (3·65 cm. high) from Brassempouy

'**The most lively artistic legacy** of the prehistoric peoples of Europe', is Hans-Georg Bandi's description of the rock paintings in the east Spanish Levant, also attributed at the time of their discovery to Pleistocene man. Unlike the cave paintings, these were found in open niches, or in small or large rock shelters, and sometimes even at the foot of a vertical belt of rock. Many of them are visible from a distance, their patina, usually ochre in colour, standing out against the grey of the stone.

The local inhabitants had long known these pictures existed. Popular belief held them to be the work of the Moors, or other foreign folk who had invaded Spain in the past. It was only in 1903, when Breuil and Cartailhac rediscovered the Altamira paintings, that the Spanish archaeologist Juan Cabré Aguilo came across some paintings near Calapata in the province of Teruel. They were executed in red, and were of an aurochs and three deer. Four years later Breuil and Cartailhac heard of this find, and they immediately began their own exploration of these rock valleys of eastern Spain, encountering there an entirely new and unexpected form of art. Discoveries came thick and fast over the next ten years; near Cogul a painting in red and black of dancing women was found, near Alpera a pictorial frieze made up of a regular medley of men and animals – men

Reconstruction drawing of the 'dancing women' of Cogul

dashing about with bows and arrows, animals rearing up at the whirr of passing arrows, a man climbing a tree to seek out honey, warriors clashing, a man with his bow drawn lying in wait in a game pass, every muscle tense. The effect was one of liveliness and movement, the figures interpreted in brilliantly simple and often almost abstract terms.

Breuil and Obermaier began by believing that the open sky and sun of eastern Spain had simply led the early inhabitants of these regions to develop a style different from that of the Aurignacians and Magdalenians living in the shelter of their caves in France and northern Spain. But others pointed out that the pictures found in eastern Spain lacked the animals typical of the Ice Age. The clothing and weapons of the men portrayed indicated that they were the product of a later epoch, when the individual was replaced by a group of people engaged in some communal activity. Herbert Kühn was one of the first to point this out. After a prolonged study of the eastern Spanish rock pictures he wrote:

A group of archers, executed in black, from the Cueva del Civil, in the Valltorta gorge, Castellón

123

'It is all so unlike Ice Age art. The style is so different that it cannot be the product of the same period. The Ice Age world is wholly real, wholly factual, firmly rooted in the contemporary. All figures, walking, hurrying, or running, are abbreviated, compressed, or altered to serve an idea, and this idea completely governs the execution of the picture.'

This view is supported not only by Hernandez Pachero, the renowned Spanish historian, but also by the man who had originally opened up this entirely new world of art, Juan Cabré Aguilo. Aguilo was born in 1882, in Calaceite in the province of Teruel. He studied in Tortosa and later at Saragossa, where he made the acquaintance of Don Sebastian Montserrat, a collector of antiquities. The two men met often and soon struck up a close friendship. This constant contact with such an experienced collector meant that the scholar became more and more drawn to the study of archaeology, although he did not neglect his marked talent for drawing. His first attempt at excavation was made at Monte de San Antonio, near his home town. Later he obtained a grant from his own province enabling him to go to Madrid, where he studied at the San Fernando Academy of Fine Arts. The discoveries he made, and his interpretations of them, soon made him one of the leading Spanish experts on prehistory.

But exactly how old were these pictures? Did they really bridge the gap between Ice Age art and the products of the Early Stone Age or even the Bronze Age, as was now supposed? Should this shadowy, silhouette-like scaling down or exaggeration of the body or various parts of the body really be seen as a sort of 'expressionism', that is, a development of the naturalistic and expressionistic style of the Magdalenian period? Hans-Georg Bandi writes:

'We must allow for the possibility that the different styles of drawing were at first merely the result of a particular intention, and that later they just became a traditional form of art. Nor can we be certain whether these distinctive methods of representation indicate various stages in the development of a style, or whether the difference between styles is merely ethnographic or anthropological. Whatever the case, it is certainly true that these interpretations of the human figure give the viewer the impression of tremendous zest, even though realism has been relatively neglected or even intentionally avoided.'

Whereas other prehistorians point out the numerous similarities these pictures have with the Ice Age pictures,

Human figures engraved in rock, Monte Pellegrino, Sicily

Bandi and others believe this affinity to be no more than a very general one. It is often emphasized that the Levantine pictures, under which category the pictures of La Brea in Sicily also belong, have striking parallels with the African rock pictures found not only in the Sahara but in East and South Africa as well. Some writers favour the idea of a 'Euro-African' hunting culture spreading in the late Ice Age from South Africa to southern Italy and eastern Spain, and which can still be seen echoed in the work of the Bushmen of the Kalahari. But this is all very uncertain. It is just as easy to imagine this to have happened the other way round, that it spread southwards from the Pyrenean peninsula into the continent of Africa. The origin of Levantine art is still a mystery, despite the close study made of it; but its end can be seen in the artistic achievements of the Mediterranean, African and Oriental Bronze Age.

A seated figure, executed in red, with a fan-shaped head, Levanzo, Sicily.

The Abbé Breuil studying the painting of the South African 'White Lady'

This great epoch of artistic achievement, which occupies so important a place in our cultural history, can no longer be studied in detail everywhere: much of it is disappearing fast. These pictures have been exposed for long to wandering herds of sheep and goats, and have suffered accordingly. What is worse, they have only too often been sprinkled with water, touched or rubbed up by visitors desiring a clearer outline for their photographs, and this has caused them serious harm; for the dampened parts attract dust, and the acid contained in the water penetrates the limestone. Added to this, foolish souvenir hunters and visitors' daubings have severely damaged the paintings. Many rock pictures are no longer visible unless they are moistened. 'If nothing is done about this', warns Hans-Georg Bandi, 'in a few years' time all visitors will be able to see is the site where once it was possible to take in the full beauty of interesting dance scenes or other paintings.'

The Grimaldi caves, near the municipality of Ventimiglia quite near Mentone, were called after the royal house of Grimaldi, from which the reigning house of Monaco also

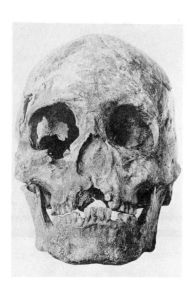

The Asselar skull (West Africa)

derives its name. In 1910 Prince Albert of Monaco founded the Institute for Human Palaeontology in Paris, engaging such men as Henri Breuil and Hugo Obermaier as professors, and arranging for systematic excavations to take place at Ventimiglia. The reason for such interest in this region was the discovery between 1872 and 1875 of the skeletons of two children, obviously interred with much love and care, in the so-called 'Grotte des Enfants'.

Later more remains of Ice Age men came to light in the Grimaldi caves; for these caves turned out to have once been Ice Age family vaults. The relatives of the departed had provided tools, jewellery and food for the journey to the next world. For a long time two of the skeletons provoked a most lively scientific discussion. They looked so different from other members of the Cro-Magnon race and other late Ice Age races found in Europe, Asia and Africa, their skulls being long and elliptical, their faces low and broad and their jaws protruding, that the French anthropologists Verneau and Rivet took them to be ancestors of the Negro.

It would be astounding, even amusing, to find that the Negro race originated in Europe. It might even do something towards shattering the racial pride of white men. However, these two divergent skulls were recently subjected to a fresh and thorough analysis, which proved them merely to have been distorted; hence their negroid look. In fact they were Cro-Magnon people, like all the others found in the Grimaldi caves. The fossilized remains of men who could possibly belong to the ancestors of the Negro have been found till now only in one or two places in Africa, for instance near Asselar in the Sahara. But even these do not differ vastly from the basic type of late Ice Age man.

Meanwhile graves and skeletons of Ice Age men have turned up in many regions of the ancient world. Some were found in abandoned habitations, others in ditches, some even under artistically arranged slabs of stone. Each body had some of its possessions buried with it, and it was in these graves that the prettiest products of Old Stone Age

Rear view of the Lespugue 'Venus',
carved from mammoth tusk

art were found: carvings, and other miniature works of art. When we consider their artistic talent and their reverent burial customs, these hunting nomadic races seem suddenly less strange, less far removed from us. Geoffrey Bibby expressed this feeling of kinship in the following words: 'Modern Europe suddenly realized that it could be proud and even a bit envious of its distant ancestors living on the borders of the ice.'

But in one respect these Old Stone Age graves do differ vastly from those of today. The dead are all laid in a particular position, their legs drawn up, like a child curled in sleep. And the bones of the feet have often been pierced. Ceremonial burial in this contracted position was still the custom in the Late Stone Age, and is practised by some primitive peoples in Oceania to this day. At first it was given a romantic interpretation: Ice Age men had laid their dead in the earth in the foetal position, that is, the position in which they had first come into the world.

But Hermann Klaatsch, a sober naturalist who does not think Old Stone Age man could have been inclined to such philosophical speculation, arrived at a more down-to-earth conclusion. Like so many tribal peoples living today, the men of the Würm Ice Age will have feared the return of their dead in the form of spirits bent on mischief. This fear caused them to bind the dead man's body and to take other precautions to prevent his undesirable return.

These conclusions brought Klaatsch up against another custom of our Ice Age forefathers, one which seems even more alien to us: the custom of human sacrifice, and the cannibalism which goes with it. Skulls have been broken open apparently to extract the brain, and human ribs are often found among the kitchen refuse of prehistoric man. This caused some to suspect that these ancient ancestors of ours had observed an incongruous assortment of customs: not only were they great artists and loving parents, they also ate their own kind. However, Klaatsch reckons this habit of theirs was not a macabre perversion. He justifies it by comparing it with the cultural customs of present-day primitive tribes:

Opposite: woman with a bison's horn, semi-relief from Laussel, Dordogne

129

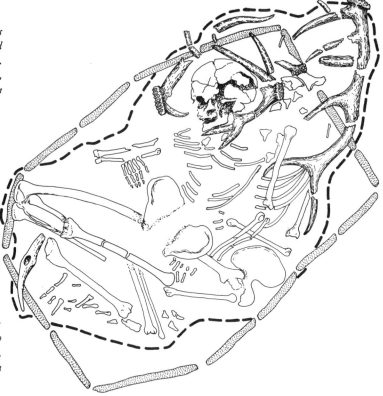

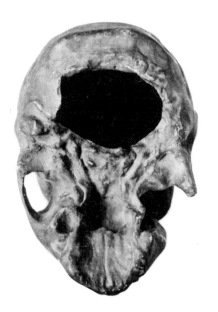

'Eating someone out of love for them does in fact have some basis in reality; they do not want to lose the one they love, and hence try to bind his soul to them by eating his body. The mother will even eat the flesh of her dead child, carrying its bones, well cleaned and dyed red, around with her. And you eat the enemy you kill mainly to absorb his strength into yourself, although the whole body may not be eaten, but only the brain, or the fat around the kidneys which oddly enough is considered a source of particular power. The idea that one can physically absorb the characteristics of another has been retained by man partly in the form of superstition, and partly in association with religious ideas.'

Thus you not only physically absorb your own dead, but also the enemy you kill so that his strength and ability may be added to yours. Many burial places reveal that even in the Old Stone Age war was an established thing among the tribes, when a stronger group of men would fall upon a weaker, destroy them, and occasionally eat their dependants too. When Klaatsch investigated some five hundred human bones discovered in a cave not far from the small Croatian town of Krapina, he went so far as to imagine a pitched battle taking place between the Aurignacians and Neandertalers.

The Battle of Krapina was long considered proof that late Ice Age man had systematically annihilated the Neandertalers. There was no sign of ceremonial burial. Hollow

The 'cannibals of Krapina', as depicted by Z. Burian

bones had been split open, presumably to extract the marrow, and some even revealed traces of fire. Beneath those who had been killed there were found the remains of a two-year-old child. However, the original discoverer, the Croatian palaeontologist Karel Gorjanović-Kramberger, re-examined the bones very thoroughly and established that there was no question of two distinct races of men warring against each other, although the bones were in such a poor condition that no one has yet been able to determine to which race they belong. In all probability Krapina was not a battleground, but some cult centre at which the local people had killed, sacrificed and eaten either members of their own tribe or, more likely, members of a rival tribe.

A prehistoric Pompeii – thus Otto Hauser, the Swiss archaeologist, designated the Vézère valley, with its caves, rock shelters, graves, stone implements, paintings and other works of art. For a long time this highly dictatorial man was not popular among other experts in his field. They disliked his aggressiveness, his exaggerated opinion of himself (a result of his ill health), and his ideas on anthropology which they considered far too Darwinian. Hauser began to excavate before the First World War, before the theory of evolution had become fully established and while it was hotly contested on all sides. Opponents of Darwinism had been happy to note that the ancestors of man so far discovered had proved to be by no means as wild and ape-like as had been feared since the publication of Darwin's theory of descent. (The more primitive and animal-like types of man which had turned up, like the Neandertal man, and the Javan early man, or *Pithecanthropus*, whom Haeckel had so splendidly prophesied, had not yet been rehabilitated.) And now this amateur arrived at the Paradise of Prehistoric Man announcing that he could bridge the gulf between the still much contested Neandertal discovery and the artists of the Ice Age by means of a series of 'connecting links'.

On a March evening in 1908 Hauser was leaving the burial site of Le Moustier when a workman told him that human remains had been discovered in a deposit as yet untouched. Because of his embattled relations with the specialists, Hauser took the precaution of leaving the deposit alone until Hermann Klaatsch happened to arrive. Together they undertook the excavation. What came to light was a skeleton with a very primitive skull, which immediately reminded Klaatsch of the Neandertal discovery. On the basis of the still imperfect methods of dating used at the time the age of Mousterian man was put at no less than 250,000 years.

A year later Hauser dug up a splendidly preserved skeleton of an Aurignacian man in the cave of Combe Capelle. The age of 100,000 years attributed to this skeleton was likewise greatly exaggerated. And then in 1910, at La Rochette, Hauser found the keystone to his reconstruction of man's evolution: some bones now attributed to the 20,000-year-old Chancelade race.

Thus systematists now had an evolutionary chart at hand which at first sight was highly attractive. It started with the

Left: Otto Hauser at Le Moustier
Right: the skull discovered at Le Moustier on 12 August, 1908

hypothetical ape-men, progressed up through the Mousterian men, the Neandertalers and the Aurignacians, to the Chancelade and Cro-Magnon races. It was not until after Hauser's death in 1932 that this so impressive-looking genealogical tree of man was again struck off the scientific record; for the Mousterian man turned out to be nothing but a young Neandertaler – merely a better preserved specimen than other known members of that type. Hauser's later finds, however, belonged in type to the late Ice Age. The age of the skeletons was substantially reduced.

Neandertal man now re-entered the picture. We owe it to the work of Gustav Schwalbe, the Strasbourg anatomist and anthropologist, that the position and importance of this prehistoric discovery came at last to be acknowledged. He made a prolonged study of all known finds of Neandertal-like men, and in 1901 proved entirely false the theory of Virchow and others that the bones were merely the remains of pathologically deformed individuals.

Schwalbe was of course able to produce far more evidence in support of his case than poor Fuhlrott, who at the time of the discovery had been forced to yield before the superior knowledge of his critics. As early as 1848 a fossilized skull

The burial of an Aurignacian man

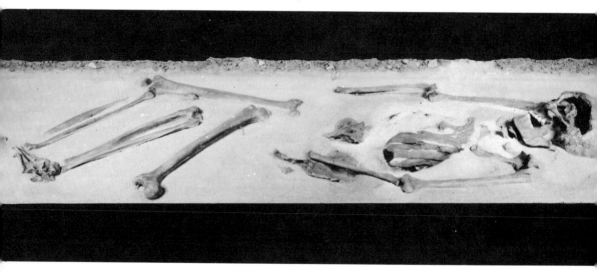

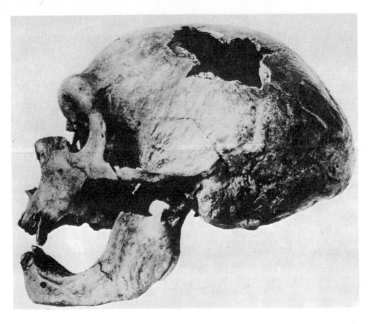

Skull of a 'classic' Neandertaler from La Chapelle-aux-Saints. Today we have the remains of more than a hundred and fifty Neandertalers, discovered in three different regions of the globe

had been discovered in excavations carried out on the north face of the rock of Gibraltar, but it was only fifty years later that English anthropologists came to compare it with the classic discovery of the Neander valley.

Then in 1866 the Belgian geologist Dupont found a human lower jaw in the cavern of La Naulette, which the anthropologist E. T. Hamy recognized as being akin to a Neandertal jaw. Sixteen years later the Czech investigator J. K. Maška came upon the lower jaw of a Neandertal child in the Shipka cave near Stramberg in Moravia. Then, between 1885 and 1886, particularly exciting finds were made by the Belgian geologists Lohest, de Puydt and Fraipont, near Namur. Here, about twelve feet down in a layer of earth as yet untouched, they dug up the skeletons of two male adults. At all these places where human remains were found not only the bones of mammoths, woolly rhinoceroses and other Ice Age creatures were present, but flint implements as well. These tools are now known to be the typical tools of Neandertal or Mousterian man.

So by the time Otto Hauser excavated the Le Moustier youth from a deliberate burial, the spell had already been broken. Shortly after Hauser's discovery three French clerics, the Abbé Bardon and the brothers Bouysonnie, discovered more bones in the small cave of La Chapelle-aux-Saints as they were digging for tools of the Mousterian period. They informed one of the leading anthropologists of France, Pierre Marcellin Boule, director of the Natural History Museum in Paris, who then uncovered a Neandertal man in first-class condition, interred in a grave about two feet down. These remains, like Hauser's find, had obviously been accorded a formal burial.

'The head of the dead man lay on a bed of stones, looking westwards, with his right hand tucked under his head as if in sleep. The legs were drawn up against his body, and over him lay the bones of reindeer and ox, probably originally intended to provide him with food.' Thus Herbert Kühn described the best-preserved Neandertal man so far discovered. His formal burial indicated that the cultural life of the Neandertalers was not essentially different from that of the men of the late Ice Age. The hole in his skull seemed to indicate a cultic cannibalism rather similar to that of the hunting nomads of the later epochs of the Stone Age. This discovery was a real surprise, for Neandertal man had hitherto been considered a primitive ape-like man – when he was allowed the privilege of being considered a man at all. Even Klaatsch, generally so very open to any new idea, at first found it hard to accept the idea of a developed Neandertal culture. He reckoned that despite these cultural beginnings Neandertal man had lived like an 'animal among animals'. It was only later, under the pressure of further evidence, that he admitted:

'We must judge the Neandertal man as a creature suited to his time, his struggles and his needs, who in many of his abilities was undoubtedly superior to us today.'

The existence of a Neandertal culture has meanwhile been proved by many further finds in Europe, Asia and South Africa. A particularly strange characteristic of these people was their cult of the bear. In the Drachenloch, or Dragon's Hole, above the Swiss town of Vättis a stone box-like construction was found covered by flat slabs of stone, inside which a number of skulls and limbs of cave bears had been prepared and arranged in a very particular way. Other Neandertal discoveries reveal that the sacrifice of the bear was an established cultic custom – as is the case with the Ainu of north Japan and some Siberian races today. Modern methods of dating tell us that the classic Neandertaler should not be regarded as the ancestor of Cro-Magnon man and other sapiens, but as a branch that went its own way while these were evolving – to give rise eventually to modern strains. Thus this early form of man, still so prominent in people's minds, and after whom a whole stage in human evolution has been named, has been eliminated from the ranks of our direct ancestors.

'The late Neandertal man was apparently well adapted to life on the cold belt of land bordering the inland ice. The heavy brow-ridges, his receding forehead, his long occipital bone extending backward, his prognathism are not in fact characteristics he inherited from his ape-like ancestors, as was first thought,' writes Professor Gerhard Heberer of Göttingen. They are special attributes which he developed as a result of his adaptation to life in a cold climate.

Robert Broom, who discovered and later described the African australopithecines, phrases it similarly: 'Neandertal man was not a very primitive man, but a highly specialized type that arose, apparently, in the late Pleistocene, perhaps as a result of the Arctic cold . . . and was not ancestral to any later form of man.'

But where did he come from? And why did he disappear so suddenly? The latter question has still not been answered. Many believe that this particular race of men did not die out naturally but was gradually eliminated by late Ice Age *Homo sapiens* with his vastly superior technique for making weapons. But the first question, the origin of Neandertal man and our direct Stone Age ancestors, can be answered at least in general outline.

It was very soon acknowledged that the ancestors of the Neandertalers as well as those of the late Ice Age races of man must be sought in much older strata than anthropologists had first assumed. In fact the relationship between these two races has only recently been sorted out. The idea of a continuous scale of human evolution, leading from the pre-Neandertalers through to *Homo sapiens*, was abolished as soon as it became clear that the development of man could not be represented by means of a straight-limbed genealogical tree, but more accurately by means of a widely ramified bush. It meant that there were many sorts of hominids living in prehistoric times, all differing very widely from one another in the extent to which they developed. Excavations of recent years have shown us which races made no advance, and which ones successfully asserted themselves.

The importance of Mount Carmel, in Palestine, is not limited to the part it has played in Jewish history. In 1931 it entered the annals of prehistoric man. Here a British anthropologist, Dorothy Garrod, had the good fortune to

Some important stages in the evolution of man, reconstructed from fossils. On this page: Dryopithecus (Proconsul), Australopithecus, Peking man. On the opposite page: Neandertal man, Mount Carmel man, Cro-Magnon man

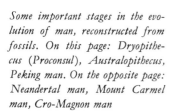

138

come across a skull as she explored a cave, a skull differing very slightly from the European Neandertal type. Extensive excavations were then carried out by British and American anthropologists under the leadership of Theodore McCown and Dorothy Garrod, not only on Mount Carmel but also around the Sea of Tiberias, the Lake of Genesareth and at Nazareth – places made famous by the parts they play in the bible story. This work brought forth the remains of eighteen prehistoric men, which, though they resembled the Neandertal type, possessed a far higher forehead and much straighter limbs. But the most surprising thing about these finds was that they proved to be quite considerably older than the European Neandertal man of the Würm Ice Age. Were they a mixed race, the result of a cross between the ancestors of the Neandertalers and those of *Homo sapiens*? Or were the Neandertalers descended from a type of man who looked far more 'modern' than they did themselves?

Other finds of Mediterranean and Asiatic Neandertalers have also turned out to be older and to look more like modern man than the Arctic bear-hunters of the last European

glaciation. And some such pre-Neandertalers have been discovered in Europe too. As early as 1871, before anyone had seriously considered trying to solve the Neandertal riddle, men quarrying in Taubach on the river Ulm in Thuringia had discovered a human skull which according to its geological deposit must have belonged to a period preceding that of the Neandertalers. Virchow declared it to be of the Late Stone Age, with the result that it was promptly forgotten. Some time afterwards human molars were discovered in the same layer of earth. They were thought to be the teeth of a chimpanzee and were also dismissed as unimportant. Then, in 1914, in the neighbouring town of Ehringsdorf, fragments of a child's skeleton were found, and in 1925 fragments of a female skeleton. It is one of the curious things in the history of anthropology that it was Hans Virchow, son of the great opponent of Darwin and Haeckel, who correctly interpreted this find.

There is no doubt that the pre-Neandertal men are linked historically with the late Neandertalers, but they lived earlier, during the last interglacial period, that is, about 150,000 years before the late Neandertal type. The pre-Neandertal strain was not nearly so robust as the typical Neandertalers of the later period; they looked more like modern man. Could this group possibly have been the common ancestor of both the late Neandertal and the Sapiens type, including modern man?

'A modern European' was the term applied by the anthropologist and geneticist Georg Glowatski to a human skull excavated some forty years ago from a gravel-pit near Steinheim on the River Murr, about thirty miles north of Stuttgart, where numerous fossilized bones of mammals had previously been unearthed. In point of fact, however, it has a distinctly archaic appearance. The dating of the skull would place it in the Mindel-Riss interglacial, between 250,000 and 300,000 years ago.

The owner of the gravel-pit, a Herr Sigrist, had over a period of some twenty years been in the habit of salvaging these bones and handing them over to the natural history collection of Württemberg, in Stuttgart. On 24 July, 1933 he reaped the ultimate reward for his industry. Below the gravel, about forty feet down, he came across the well-preserved and considerably fossilized skull of a young woman. The Chief Curator of the Stuttgart collection, F. Berckhemer, immediately sent to the site his technical assistant, Herr Böck, who managed to free the skull from the gravel adhering to it. He first took it to be of the pre-Neandertal type; but very soon this verdict raised doubts. True, the powerful brow-ridges, the broad nose and the tiny brain-case of only 1150 c.c. made it look very ancient (the Neandertal man's brain capacity of 1350–1720 c.c. exceeded even that of modern man, which ranges from 1350–1500 c.c.). On the other hand, the Steinheim skull was longer, the back of the cranium more rounded, while the face protruded far less than that of the Neandertal type. The anthropologist Hans Weinert pointed out that while he resembled Asiatic early man, or even the 'ape-men' whom we shall later describe, he also seemed to represent a precursor of *Homo sapiens*.

One of the strongest advocates of the theory that this extremely ancient and yet so modern-looking Steinheim skull held a unique position in the direct ancestry of the late Ice Age peoples and thus of modern man was the Göttingen anthropologist, Gerhard Heberer. He was also the first to coin the term 'Pre-sapiens man'. Admittedly, a single find is hardly enough to support an entirely new theory. Very soon, however, the prehistoric Steinheim woman proved not to be unique after all; she was found to have what were clearly close relatives in various regions of the earth.

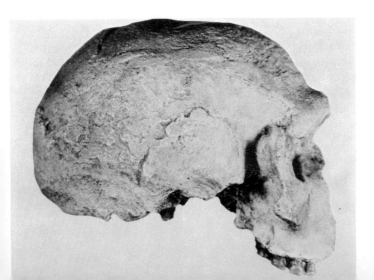

The Steinheim skull, so ancient yet so modern-looking

Swanscombe man on the hunt

Now another gravel-pit, this time at Barnfield, near Swanscombe, about thirty miles south-east of London, enters the picture. Here a London dentist, Alvan Theophilus Marston, had for many years been collecting animal bones and implements of the Acheulean age, that is, the pre-Neandertal period. In June 1935 Marston had the great good luck to discover a human occipital bone about eighteen feet down, and then in March 1936 only seven yards from this spot, and in the same stratum, the left parietal bone belonging to it. Both bones were in almost perfect condition and fitted admirably together. They too had probably belonged to a woman about twenty years old. Then in 1955 the right parietal bone turned up. As at Steinheim, the deposit which produced these bones indicated that the remains had originated in the Mindel-Riss interglacial. Despite the

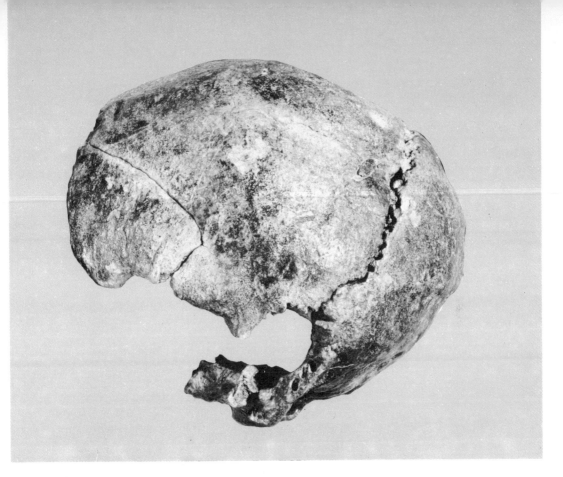

The various pieces of the Swanscombe skull were found over a period of twenty years

Hand-axes used by the Acheulean huntsmen, found near Swanscombe (about two-thirds actual size)

143

Spear-heads from various periods of the Old Stone Age, from the Aurignacian (opposite, extreme right), to the late Magdalenian (opposite, extreme left). Below: spear-head made of yew. Most spear-heads were fashioned out of reindeer antler, which was straightened while the bone was fresh; in the course of time, however, they assumed their original bent shape

breadth of the occipital bone and the robust bone structure, this Swanscombe woman bore a surprisingly close resemblance to the *Homo sapiens* type.

Further discoveries made at Quinzano near Verona and Fontechevade in central France are now also considered to belong to the pre-sapiens type. They stem from the Riss glaciation, and are consequently more recent than the Steinheim and Swanscombe finds, though still more ancient than the Neandertal. They show a closer resemblance to modern man. The British anthropologist Dr G. M. Morant remarked back in the thirties: 'It appears likely that the group they represent was either in the direct line of descent of *Homo sapiens*, or if not in the direct line at least closer to it than was H. neanderthalensis.'

Pre-sapiens man was not confined to Europe. Skulls found on Mount Carmel reveal by their breadth of face and prominent chin that they belonged to beings who were at least closely connected to the pre-sapiens race, and perhaps, as Glowatski says, they are even 'a mixed race descended from the pre-Neandertalers and pre-sapiens men'. In 1932 the palaeontologist Louis S.B. Leakey, who has since made a name for himself as the discoverer of East African early man, came across four very old but sapiens-like skulls in East Africa, near Kanjera in Kenya. Another cranium discovered in South Africa, though still the subject of controversy, appears to belong to the same group. It was found at Florisbad near Bloemfontein in 1932, in a cave on the banks of a pool near a mineral spring deposit, and despite its massive forehead looks less like the Neandertal than the European pre-sapiens type, notwithstanding the presence of a few more primitive characteristics.

One very surprising development has been the discovery in various regions of cylindrical spearheads, and even entire wooden spears, found among the numerous tools of the Acheulean period – in the 'Elephant Bed' of Clacton-on-Sea, for instance. These are obviously the weapons of pre-sapiens man, and even though more advanced they are definitely older than those of the pre-Neandertal man. Thus the question which anthropologists now face is this: Why

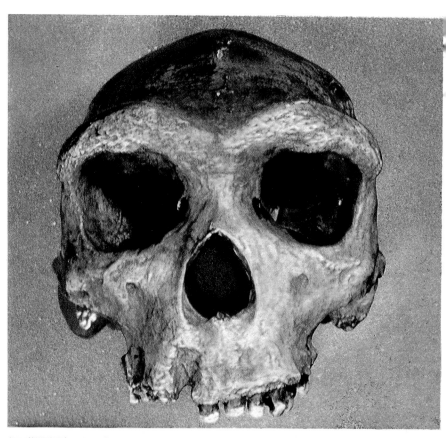

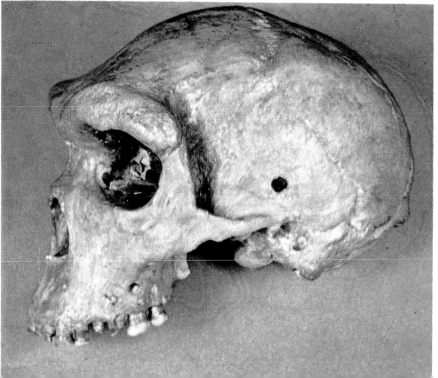

should we suppose the sapiens people to have developed out *Opposite: The Broken Hill skull*
of the Neandertal or pre-Neandertal when sapiens-like types
existed long before? Most experts today believe that pre-
sapiens man was at least a contemporary, if not an 'elder
brother', of the pre-Neandertal type.

This, in broad outline, is the history of *Homo sapiens*, the
species to which we belong. On the one hand we have the
pre-sapiens type producing the pre-Neandertal race, which
gave rise to the Neandertal who then unaccountably died
out; and on the other, the various late Ice Age races, which
had also developed out of the pre-sapiens type but which
went on to produce the ancestors of man today. But long
before these conclusions were reached a quite different and
most confusing fossil find had been enthroned as the 'oldest
European', even the 'first-born', the real ancestor of man.
This was the notorious 'Dawn Man', *Eoanthropus dawsoni* of
Piltdown.

Part Three

The dawn of humanity

'Man is the measure of all things;' this
statement also holds true for the history of
evolution. Getting to know the history of ammonites,
cuttlefish, horses or elephants is all very fine
and interesting, but there is nothing more important
to us than the study of our own development.

G.H.R. von KOENIGSWALD

DAWSON

SMITH-WOODWARD

OAKLEY

KNOPF

LIBBY

SCHOETENSACK

DUBOIS

VIRCHOW

von KOENIGSWALD

OPPENOORTH

TEILHARD DE CHARDIN

ANDREWS

BLACK

PEI WENG CHUNG

WEIDENREICH

KOHL-LARSEN

'On the staircase of the Geological Society in London
there hung a large picture, probably painted shortly before
the First World War. Sitting behind a long table are to be
seen three earnest looking men whom the initiated could
easily recognize as England's most celebrated anthropologist Sir Arthur Keith, the craniologist Professor Elliott
Smith and the palaeontologist Sir Arthur Smith-Woodward.
Grouped around them was the entire staff of the British
Museum, and in their midst, modest and self-effacing, stood
Mr Charles Dawson. The object which had brought him
sudden world fame lay on the table before them: the cranial
remains of *Eoanthropus*, species *dawsoni*, so named in honour
of its discoverer.'

Thus von Koenigswald recalls the time when the Piltdown
man was still considered by many highly esteemed scientists
to be the showpiece in the demonstration of man's descent,
the positively ideal missing link in the chain of human
development – in short, the real Adam. The one who held
this view most strongly was Sir Arthur Keith, the Nestor of
British anthropology. Although many experts pointed out

shortly after the discovery that a creature possessing such a well developed human skull and yet so ape-like a jaw could hardly be fitted into Darwin's picture of human evolution and could therefore be no more than a freak of nature, Sir Arthur persisted in trying to fit the ape's jaw somehow into the human skull. And to the end of his life he held firmly to his conviction that the skull and jaw belonged to the same creature and that together they formed an absolutely ideal 'ape-man'.

A motion of truly national importance was put before the Speaker of the House of Commons by six Members of Parliament on 26 November, 1954: 'That the House has no confidence in the Trustees of the British Museum . . . because of the tardiness of their discovery that the skull of the Piltdown man is a partial fake.'

The reason why the matter seemed of such national importance to those who proposed the motion was that the Trustees of this the greatest museum in the world included such eminent people as the Prime Minister of the day, Winston Churchill, Foreign Minister Anthony Eden, the Archbishop of Canterbury, and even a member of the royal family. This vote of no confidence was aimed to publicize the fact that for forty years British science had been hoodwinked by one of the cleverest forgeries ever perpetrated in the course of man's study of his prehistoric past.

The reaction of the Speaker of the House of Commons was to declare: 'I should like to consider it. Speaking for my statutory co-trustees, the Archbishop of Canterbury and the Lord Chancellor, I am sure that they, like myself, have many other things to do besides examining the authenticity of a lot of old bones.' But despite this levity the Piltdown case became a scientific scandal, amusing some but deeply shocking others, not only in England but throughout the world.

What had happened? In 1908 Charles Dawson, lawyer and amateur anthropologist, a quiet and diffident man, produced a brown cranium found in a gravel ditch near Piltdown in Sussex. He had apparently been given it by a road-worker.

'Searching for the Piltdown Man'. This postcard, which appeared around the time of the discovery, shows Charles Dawson (left) and Dr Smith-Woodward (right) busily digging

And, according to his account, he had found more skull bones himself over the next few years, some in the gravel of the road and some in a ditch bordering a field. But it was only in 1912 that he got in touch with Sir Arthur Smith-Woodward of the British Museum. Together they explored the site, and found a very ape-like jaw and a few teeth. Smith-Woodward was surprised at the unique colouring of the bones; and when Dawson explained he had treated them with potassium dichromate to harden them, Smith-Woodward told him he had been wrong to do so. The fact escaped him that the ape-like jaw which he himself 'found' was similar in colour.

Critics soon pointed out that the lower jaw must belong to an ape whereas the skull was human; for despite all Sir Arthur Keith's attempts at reconstruction the two just did not fit together. But, as Ruth Moore says, one scientist who reckoned the Piltdown find to be authentic commented drily

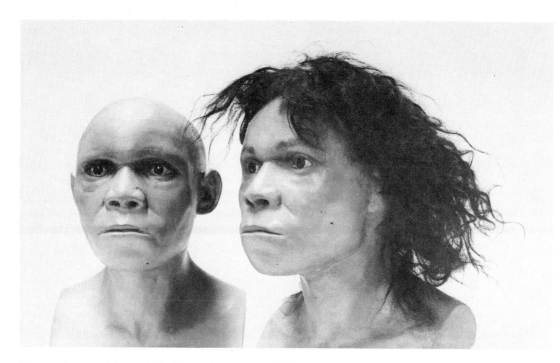

In 1951, the year of the Festival of Britain, many anthropologists still believed in Piltdown man, as these reconstruction busts by Maurice Wilson show

that it would be a miracle if a primitive man left his brain-case and not his jaw in the very spot where an anthropoid had left his jaw and not his brain-case.

The value Smith-Woodward attached to this 'Dawn Man' is described by von Koenigswald in his memoirs: 'He had a small house built for himself at Haywards Heath, not far from the site of the find, so that he could always keep an eye on it . . . From then on he dedicated his whole life to Piltdown man. When we visited him at Haywards Heath in 1937 he talked of nothing else . . . Standing under a big umbrella, Sir Arthur showed us the spot at which he had unearthed the celebrated find.'

Most anthropologists were not at all happy with the way the discussion about the Piltdown miracle was developing. Although the French theologian and palaeontologist Pierre Teilhard de Chardin himself uncovered another ape-like tooth at the site, the monster these bones suggested, with its miraculously capacious cranium and really chimpanzee-like jaw, did not fit into the theory of human evolution which had by then begun to be accepted and which is now established. This theory holds that the skull of prehistoric man developed little, while the jaw and teeth acquired comparatively human features. With the Piltdown man it was precisely the reverse.

What is more, no further finds having been made anywhere near the vicinity of Piltdown since Charles Dawson's death in 1916, there was by the thirties a growing suspicion among certain anthropologists that the skull was no more than a cunning forgery.

By 1949, like a squad of detectives, the Englishmen Kenneth Oakley, J.S. Weiner and Sir Wilfred Le Gros Clark were circumspectly collecting evidence towards demolishing the claims of this troublesome fossil. More sophisticated methods had meanwhile been developed for dating fossilized bones, either by determining their nitrogen content, or by the fluorine test which Oakley had introduced. Bones lying in the ground absorb the fluorine contained in small quantities in the ground-waters. The longer they lie there, the more fluorine they absorb. Gerhard Heberer describes the tests these three men carried out on the Piltdown bones: 'An accurate examination of the lower jawbone and a perfected method of testing the fluorine content revealed that this ape-like jawbone was of recent origin and had been artificially and fraudulently made to resemble the condition of the human brain-case.' Oakley could even prove that the teeth had been skilfully 'fossilized': the master forger had introduced tiny grains of ironstone into the nerve canals.

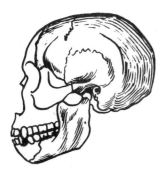 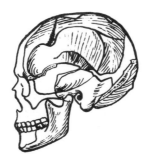

Like a jig-saw puzzle the faked Piltdown bones were fitted together again and again, and widely differing versions were produced.
Left, Dr Smith-Woodward's reconstruction; right, that of Professor Keith

153

Who could the forger have been? As in so many cases of crime committed some fifty years before, this is something about which we cannot be certain. Nevertheless, Professor von Koenigswald writes of Charles Dawson: 'It is certainly not nice to accuse a dead man who cannot defend himself; but everything quite clearly points to his responsibility for the forgery. Indeed, it has now turned out that neither the fossils nor the tools belong to this locality at all, and that the whole find was carefully planted.'

In a final bulletin about the affair Sir Gavin de Beer, then director of the British Museum, closed the case: they had laid the ghost of Piltdown man, who, to be honest, never fitted very happily into any evolutionary scheme. In fact they had even won some advantage out of it after all; for it tested half a dozen new experimental procedures for the investigation of fossils which had made the recurrence of such an event impossible, and so had considerably forwarded the scientific study of fossils. And Oakley exclaimed tongue-in-cheek about the mistake they had made: '*Of course* we believed that the great brain came first! We assumed the first man was an Englishman!'

Kenneth Oakley (left) discusses the ape-like lower jaw found near Piltdown, with E.L. Parsons, in the British Museum (Natural History)

Another find as much disputed as Piltdown man was the discovery made by workers at a zinc mine at Broken Hill in Zambia, formerly Northern Rhodesia. As they were demolishing a hill about sixty feet high they came across a blocked-up passageway which they found ended in a cave. Here they probably found many more human bones, perhaps even whole skeletons, than the few that reached the scientists' hands; among these was a great capacious skull with a receding forehead, powerful brow-ridges and 'a surprisingly brutal facial structure', as the report describes it. The bones were obviously ancient.

However, Rhodesian man was not a forgery, although many of the conclusions first drawn from his discovery turned out to be false. Unfortunately the first study made of the skull was not a careful one. The circumstances in which it had been found were positively confusing, the geological level no longer ascertainable, and although the British Museum received the skull as early as 1921, seven years elapsed before a detailed – though even then still incomplete – report actually appeared.

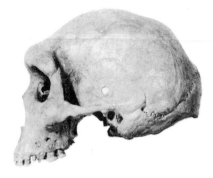

The skull of Rhodesia man (Homo erectus rhodesiensis), side view.

It is true that for a long time the Broken Hill discovery took second place to the Piltdown man, even though some scholars believed it to exceed all others in age. Even Klaatsch toyed for a while with this idea: 'This great male skull looks far more ancient than do many Neandertalers. Therefore it is quite understandable that Rhodesia man was taken to be a direct descendant of the gorilla. But, alongside these primitive characteristics there are also details, particularly in the face, which resemble those of *Homo sapiens*.'

But other doubts arose as well. Palaeontologists pointed out that this remarkable creature of Broken Hill had undoubtedly suffered from tooth decay. It was difficult to imagine how this disease of civilization could have attacked prehistoric man. And two very odd holes in the side of the skull caused the experts even greater perplexity. In the view of Professor Mair of Berlin they looked like the entry and exit holes of a modern bullet. Could Rhodesia man be some sort of 'surviving fossil' which modern huntsmen had despatched with a bullet in the brain?

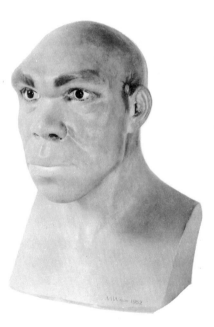

Maurice Wilson's reconstruction bust of Rhodesia man

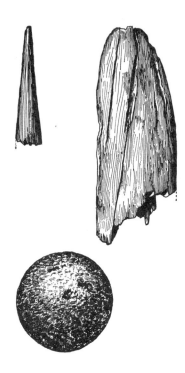

Implements of Rhodesia man: above left, a bone point, or awl: above right, a bone gouge. Below, a spherical missile made of granite

Broken Hill man is still a mystery today. According to the modern methods of dating which we will describe below, he is no older than the European pre-Neandertalers and a fraction younger than representatives of the pre-sapiens type. He is often wrongly called the 'African Neandertal man', as is the Saldhana man later discovered in South Africa and who in Oakley's view is somewhat older. Other scientists, on the other hand, consider this primitive African to be a late form of the early species of man once known as *Pithecanthropus* and now as *Homo erectus*, and which had apparently died out (see pp. 167–179). As for the mystery of the holes in the skull, this still remains unsolved, although the theory has been advanced that a prehistoric hyena had sunk its teeth into the skull.

'Clocks of the earth' is a term applied to those modern dating methods by which the age and length of each epoch of the earth's history and its fossils are determined. In the early days the fluorine and nitrogen content had been enough to expose the Piltdown forgery. But now there are many other far more advanced physico-chemical methods for establishing when a creature whose fossilized remains are found inhabited the earth. A select Committee was formed in the United States in 1929 to study the problem of the age of the earth and the various methods by which this could be measured. The Yale Professor Adolph Knopf, chairman of this committee, hit upon the promising idea of using the decay of atoms as a means of measuring geological time.

All radioactive elements, such as Carbon 14 and the uranium isotopes 238 and 235 are unstable, breaking down in one or more stages, until a stable element is reached. The rate of breakdown varies according to the isotope, being of the order of a few thousand years for Carbon 14, and millions of years for uranium.

Uranium minerals are present throughout the earth's crust. The time taken for these minerals to disintegrate is known precisely. Each stage results in a certain quantity of lead and helium. If the ratio of the substance produced by decay is measured against the ratio of uranium still present, this tells us how long the process of disintegration has been going on and hence the age of the material under observation.

Apparatus for determining age by the radiocarbon method at the University of Cambridge

157

'Dating the Past' was the title of a book by Frederick E. Zeuner, published in 1946; in it he not only discussed the uranium clock but other 'clocks' of the earth as well. If we call the uranium method (with its slow breakdown rate) the hour hand of the earth clock, then in the absence of an available isotope, we can think of relative dating by the nitrogen, fluorine and iron content as the minute hand. The hand ticking off the seconds was finally discovered by the American scientist, Willard Libby, who found that all organic nature contains not only the normal carbon C^{12}, but also the radioactive carbon C^{14}, produced in the upper atmosphere by cosmic ray bombardment of nitrogen atoms.

This radioactive carbon isotope mixes with the carbon dioxide of the air, is absorbed by plants, and then transmitted to all organisms; for the plants are eaten by animals, they in turn are eaten by other animals, and of course man eats both plants and animals.

As long as an organism is alive, whether animal or plant, the loss of C^{14} as a result of atom decay is renewed by the fresh intake of C^{14} with food. It is only at death that the radiocarbon clock begins to run. Slowly and inexorably, as

Professor Willard F. Libby in the date-testing annexe at the University of California, Los Angeles

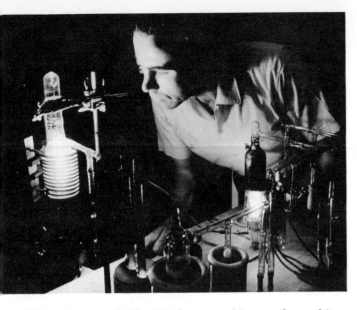

The *potassium-argon method is the most recent device for age determination*

steadily as clockwork, the C^{14} decays and is transformed into normal carbon C^{12}. Once Libby had discovered how this decay takes place all that was needed was the physical apparatus capable of making an accurate measurement down to the tiniest trace of the presence of C^{14}.

Another application of the 'time clock' also based on nuclear processes but more adaptable than the C^{14} method, has recently been devised: the potassium-argon method. It promises to fill the gap between uranium and carbon isotope dating; for by this means it is possible to determine the isotopes potassium 40 and argon 40 in geological strata older than those to which the C^{14} test can be applied. Thus it is now theoretically possible to establish the exact age of those near-human and early human beings which link us with the animal world. This test has also shown us that the Pleistocene age, that is, the Ice Age, began two or three million years ago, not a mere million years ago as had earlier been supposed. Back in 1952 a colleague of Leakey's, Frederick Johnson, was so optimistic as to declare that all these timepieces of the earth would eventually make it possible for us to divide world history into chronological epochs with complete accuracy.

The memorial stone near Mauer

In the cemetery of Mauer, near Heidelberg, is the grave of a modest quarryman named Daniel Hartmann. Hartmann is today an honorary citizen of his small home town. Not far from here, near the road linking Heidelberg to Heilbronn, the visitor's attention is caught by a singular and highly unusual memorial stone upon which a human lower jaw is depicted. Its inscription reads: Homo Heidelbergensis 21.10.1907. When this massive-looking jaw-bone with its powerfully broad condyles was discovered, Hermann Klaatsch passed a verdict which at the time seemed very sensational: 'The thoroughly crude and downright animal look of the Mauer mandible can have no connexion with the Neandertal type, even assuming an earlier stage in the development of this race of man.'

In other words, soon after his discovery Heidelberg man seemed as if he would topple the ill-fated Piltdown man from his throne. Many were convinced he must be the oldest European, far older than Neandertal man who had till then been in the running for this position, and much older too than pre-sapiens man, who was discovered much later and is now in great favour with anthropologists.

Unfortunately only this single lower jaw has so far come to light. Its discovery was again due to the care and alertness of a gravel-pit owner, the sharp eyes of a workman and the unwavering perseverance of a modest and therefore all the more congenial scientist. The gravel pit at Grafenrain near Mauer had already provided Heidelberg University with a real wealth of fossils of the Günz-Mindel interglacial of 500,000 years ago. Its owner, J. Rösch, was in regular contact with Otto Schoetensack, geologist at Heidelberg University, to ensure that not so much as a single bone, however small, would be lost by the men working the pit.

On 21 October, 1907, fifty-two-year-old Daniel Hartmann thrust his shovel into a layer of earth eighty-two feet from the surface and brought up a bone which broke in two as he lifted it. Hartmann immediately realized it might be a human lower jaw and he sought out Herr Rösch, who, after a glance at the find, dashed off a message to Otto Schoetensack: 'For twenty long years you have sought some trace of early man in my pit . . . yesterday we found it. A lower jaw belonging to early man has been found on the floor of the pit, in a very good state of preservation.'

Schoetensack describes the subsequent events: 'The next train took me to Mauer, where I found the information I had received fully confirmed. The jaw was broken in two, but the halves had still been together when the workman's shovel struck it in the earth. It fell apart as it was lifted out. Around it and on the incisors and molars was a crust of rather coarse sand, as had been the case with the animal bones already dug up from the Mauer sand.'

Within a year Schoetensack had published an exemplary description of the discovery and also of the Mauer fauna. In it he points with great accuracy to the special characteristics of the Mauer mandible, characteristics which have so racked the brains of experts ever since. The jaw could have belonged to a prehistoric anthropoid ape, in that it lacks a protruding chin and the projecting ridges for speech muscles. But the teeth, on the other hand, are just like those of modern man. Hence some anthropologists hit upon the rather grotesque idea that this strange creature could perhaps have been a giant gibbon with human teeth. But others were less interested in the shape of the mandible than in its dentition: this ancient being dug up out of Herr Rösch's gravel pit was in their view a primitive *Homo sapiens*, despite its great age.

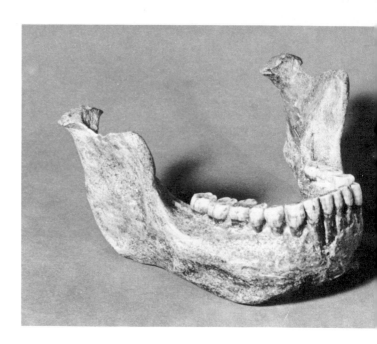

The lower jaw of Heidelberg man (Homo erectus heidelbergensis), found in the gravel pit near Mauer

'The teeth are typically human; the canines do not project above the level of the other teeth, and the third molar, which in primitive races of men – for instance often in the aboriginal Australians – is similar in size to or even larger than the second, is smaller in the Heidelberg jaw, just as in our more advanced races today . . . This Heidelberg jaw is quite definitely a fragment of true man, a representative of the species *Homo sapiens.*' This verdict was pronounced by Johannes Ranke, a stubborn opponent of the theory of descent. He supported the anti-Darwinists, maintaining that the Mauer mandible proved human culture extended 'at least as far back as the oldest Tertiary', and that Tertiary man had already possessed 'the essential physical characteristics of modern man'. In other words, any search for the missing link was really unnecessary, for it was not the ape but man himself who was in fact born first. His great age made it impossible for him to have arisen out of any sort of ape.

In fact the Mauer mandible has absolutely no connection with the Tertiary, let alone the earliest Tertiary, as Ranke had assumed. Schoetensack had known this right from the start, and it has been confirmed by the painstaking investigations carried out by the many scholars who have since followed in his footsteps. Heidelberg man lived 500,000 years ago, towards the end of the last interglacial period, when Europe was enjoying a really temperate climate with all its typical flora and fauna. Thus he is at least as old as the oldest Asiatic early man whom we shall soon describe, and with whom he is linked by most anthropologists today. We first learned something of his way of life from discoveries made in Hungary between 1964 and 1965.

'The earliest hearths of humanity' is how the Hungarian archaeologist László Vértes described an Old Stone Age camp site estimated to be around 400,000 years old which he discovered near Vértesszöllös in western Hungary. Vértes

Excavation at Vértesszöllös

163

noted in his first report that the deposit which produced this find is about 5 cm. thick, and that in it he found the bones of sabre-toothed tigers, early rhinoceroses and beaver. The stone implements looked far more primitive than those of the Acheulean period, the time of pre-sapiens man. The method used by these even older Europeans was to strike one pebble with another until it had been chipped into the required shape. What was particularly interesting about this Hungarian discovery was that most of the bones revealed traces of fire, and that there were indications of regular hearths in the same deposit. Thus man had already pre-empted the legendary Prometheus: he had discovered the use of fire.

Skull fragments found at Vértess-zöllös; it is reasonable to suppose that they are related to the Heidel-berg find

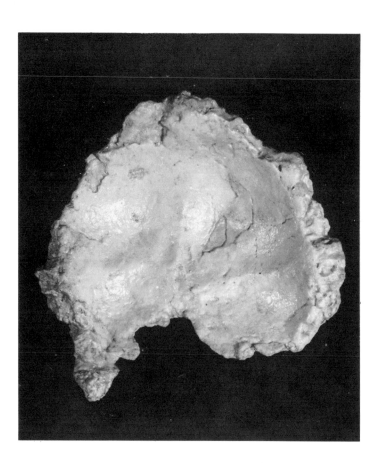

These regular hearths and the great number of bones lying about led to the natural conclusion that such places had already taken on the character of settlements. Then in 1965 Vértes came upon a fragment of a human skull and a few isolated human teeth in the same layer of earth. That there were affinities with the Mauer jaw-bone was immediately obvious. What is more, the settlement, the heaps of bones and the regular fire places had a marked similarity to places long known to exist in east Asia. Could the first Europeans have possibly been immigrants from the Far East? Was Asia therefore the cradle of humanity? This was an idea which had been broached in antiquity. It was to occupy men's minds afresh as the result of the exciting find of an early human ancestor in Java.

'P.e.175mONO 1891/93' is the apparently puzzling inscription upon a stone set by the side of the Solo River near Trinil in central Java. It marks one of the greatest and at the same time most controversial discoveries ever made, a discovery which has intrigued palaeontologists and archaeologists for nearly fifty years and has caused one of the most violent controversies ever to take place in connection with the study of human origins. The inscription means that in 1891–3, 175 metres east-north-east of that particular spot, *Pithecanthropus erectus*, the much famed but also much maligned 'Erect ape-man of Java' was found.

Von Koenigswald describes the importance of this discovery as follows: 'One cannot open a single book on anthropology, prehistory or the comparative anatomy of man, palaeontology or geology, without at least finding his name. A primitive man or a specialized ape? No other palaeontological discovery has created such a sensation and led to such a variety of conflicting opinions. It was this find that brought the problem of the coming of man before the public.'

This was roughly the time when Ernst Haeckel, in Jena, was postulating his *Pithecanthropus alalus*, or 'speechless ape-man'; but it was also the time when Rudolf Virchow was adamantly refusing to accept the early human remains so far

The memorial stone near Trinil in central Java, erected beside the Solo River

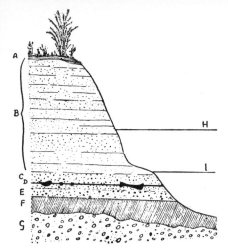

Eugène Dubois' drawing of the excavation site at Trinil. The finds came from level D; H represents the high-water mark during the rainy season

discovered as anything more than pathological individuals. The vast majority of ordinary people moreover still treated the idea of the missing link, of a link between apes and men, as ridiculous, even blasphemous.

Eugène Dubois, a young Dutchman, took up the post of military surgeon in the Dutch East Indies for the express purpose of using his free time there to hunt for primitive men. Dubois had been fired by Haeckel's convictions. As early as 1888 he had written a long article about the Indonesian fauna in which he says: 'Since all apes, and notably the anthropoid apes, inhabit the tropics, and since the ancestors of man as they gradually lost their coat of hair will most certainly have lived in the warm areas of the earth, we are forced to conclude that the tropics is the region where we should expect to find the fossil ancestors of man.'

'Looking for a needle in a haystack' was the way people later described this young military doctor's pursuit of a phantom. His attempts at excavation – first on Sumatra and then on Java – had in fact the oddly erratic character of a dream-sequence. In Wadjak in central Java he discovered two prehistoric-looking human skulls; but their large braincases indicated that they were south-east Asian relatives of European Ice Age man. Dubois took these Wadjak skulls to be the ancestors of the aboriginal Australians or Papuans, packed them away and said nothing about them till 1922. Neither did he mention a very badly preserved fragment of a human mandible which he dug up in November 1890 near Kedung-Brubus in the Solo valley, and which we now know to belong to the same type of man as his great Trinil find.

In September of the following year he found a right upper molar in a river cave. Although not yet sure whether it was the tooth of an ape or of early man, he enlisted the services of all the Malays he could muster and made a systematic excavation of the 'Trinil Hell', as the Dutch then called the Solo region. This uncanny man seems almost to have gone about his work of excavation like a clairvoyant in a trance, somehow doing exactly what was needed for the realization of his dream.

'From the inventor of Pithecanthropus to his happy discoverer!' Ernst Haeckel telegraphed Eugène Dubois immediately he heard of the latter's great find. The extensive excavations carried out by Dubois on the Solo River had produced a skull-cap which looked like that of an ape with large brow-ridges, a left upper molar, and then a much-fossilized human thigh-bone. Despite his ardent belief in Haeckel's ape-man, Dubois did not yet dare to relate his three finds to it. The skull-cap could well belong to a chimpanzee-like ape, the thigh-bone, on the other hand, to a man. Not until 1894 did Dubois take the bull by the horns and, in an exact description of his find, announce: '*Pithecanthropus erectus* is the transitional form which according to the theory of descent must have existed between man and ape. He is the progenitor of man.'

Haeckel was overjoyed: 'Now the state of affairs in this great battle for truth has been radically altered. Eugène Dubois' discovery of the fossil *Pithecanthropus erectus* has actually provided us with the bones of the ape-man I had postulated. This find is more important to anthropology than the much-lauded discovery of the X-ray was to physics.'

The famous skull-cap of the controversial 'ape-man' found at the Solo River site (left lateral view)

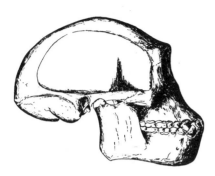

The first reconstruction of the Pithecanthropus skull, made by Dubois in 1896

Thus, at a stroke, did Pithecanthropus become the focus of world attention. On 14 December, 1895, the Berlin Society for Anthropology, Ethnology and Prehistory arranged a special gathering of its members to discuss this 'ape-man of Java'. Experts came from all over the world to inspect the skull, whose brain capacity was estimated to be between 800 c.c. and 1000 c.c., that is, exactly between that of an ape and that of a modern man. They inspected the thigh-bone, which lacked any sign of the bent posture of the Neandertalers and looked far more like that of modern man. This ape-like prehistoric ancestor of ours had not walked in the manner of an ape: he had walked erect, a true Adam, and was registered as such by his specific name of *erectus*.

Virchow would not take the chair at the Berlin meeting. He listened attentively to the lively debate, which at times provoked a virtual tumult, and heard the Swiss anatomist and anthropologist Kollman describe Pithecanthropus as a primeval ape which 'had reached the borders of variability and had become a lasting type'. He listened to the answer of a colleague arguing that the progenitors of man had not been great apes but smaller varieties, out of which pygmies and later the larger human races had sprung. He shook his head as the zoologist Alfred Nehring endorsed Eugène Dubois' ideas. Then finally, with the trained eye of a pathologist, he pointed to the irregular growth on the thigh-bone: 'I know such growths, and have treated many patients who had them. If these patients had not been well looked after they would have died. But the creature who possessed this dubious thigh-bone did not die as a result of his complaint; the growth healed, as we can see. Thus it cannot be the bone of a primitive man but of a modern one.'

When his opponents pointed to the skull-cap as proof of their thesis Virchow was still unconvinced. 'The skull has a deep suture between the low vault and the upper edge of the orbits. Such a suture is only found in apes, not in man. Thus the skull must belong to an ape. In my opinion this creature·was an animal, a giant gibbon, in fact. The thigh-bone has not the slightest connection with the skull.'

Similar debates took place in the Dutch university town of Leiden, in Paris, London, and various other places where Dubois displayed his finds.

But democratic procedures were allowed to prevail. The participants at the Berlin and Leiden congresses voted in parliamentary fashion on the nature of Pithecanthropus. The Berlin zoologist Wilhelm Dames collected the statements of twenty-five scholars. Three opted for the decisive word 'ape', five for the definition 'man', six for the 'missing link', a further six for a creature 'part ape, part man', and two for 'a link between the missing link and man' (a sagacious verdict commanding our utmost respect).

Virchow, however, stuck to his guns: 'All I can do is warn against drawing decisive conclusions from these few pieces of bone about the greatest question facing us in the study of our creation. Pithecanthropus will remain doubtful as a transitional form until someone can demonstrate how this transition, which to me is conceivable only in my dreams, actually came about.'

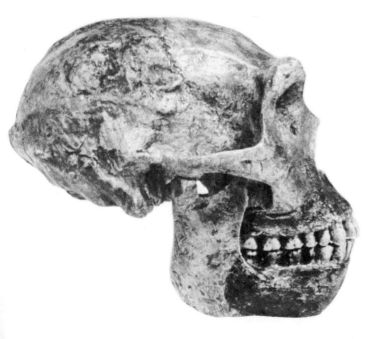

The skull of Pithecanthropus modjokertensis from Sangiran in central Java. Recent reconstruction model by G. H. R. von Koenigswald

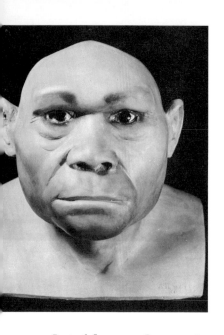

Bust of Java man. Reconstruction model by Maurice Wilson

Wilhelm Dames, chronicler of this debate, summed it up as follows: 'Great differences of opinion usually result in uncertainty and vacillation. In this case, however, they can actually be considered to have lent strong support to the transitional nature of Pithecanthropus.' This verdict was supported not only by Haeckel but by such well-known anthropologists as Klaatsch and Schwalbe; whereas Ranke, adamant anti-Darwinist as he was, reckoned the Trinil find to be a great man-like ape 'which even text books quite falsely proclaim as the transitional form between man and ape we have sought so long in vain.'

Pithecanthropus became Dubois' destiny. Instead of being pleased about the differences of opinion expressed about his find he was resentful, and from then on kept the bones he had displayed at the congresses firmly locked away. Ernst Haeckel, 'inventor' of Pithecanthropus, who had not been present at the Berlin and Leiden congresses but who enthuses about the find in many of his books, never even managed to see the bones. What is more, Dubois did not so much as mention that he had the remains of not less than four other thigh-bones locked away in his boxes. Von Koenigswald, who visited him in 1936 and was one of the first scientists to be granted another look at the mystery-ridden Pithecanthropus bones, wrote of Dubois, now living a solitary existence in the Dutch town of Haarlem:

'Pithecanthropus was his discovery, his creation, his exclusive possession; on this point he was as unaccountable as a jealous lover. Anyone who disagreed with his interpretation of Pithecanthropus was his personal enemy. When his ideas failed to win general acceptance he sullenly withdrew, growing mistrustful, unsociable and eccentric. He showed his finds to hardly anyone, and at night he was sure he heard burglars prowling around the house bent on stealing his Pithecanthropus. If anyone came to the house in whom he scented a colleague he was simply not at home.'

It is even said that this strange man later completely rejected his earlier interpretation of his finds, for what reason we do not know. When von Koenigswald, at

scarcely thirty years of age, went to Indonesia to find conclusive evidence of Javan early man, Dubois is reported to
have asserted that his 'ape-man' was nothing but an old
giant gibbon after all.

'What use is such a discovery if it is buried all over
again?' Dubois is reported on one occasion to have snapped
at the experts opposing him. Although he himself was partly
responsible for this second burial of Pithecanthropus, he
awaited the results of the new expedition in some suspense,
for it amounted to a continuation of his own work in Java.
One of the most prominent authorities on the south-east
Asian animal world at the turn of the century was the Munich
zoologist, Emil Selenka. He was regarded as one of the
leading experts in the study of men and apes and merely on
this account had a deep interest in human origins. So he
organized a German-Dutch expedition, which was to make
a thorough investigation of the site in the Solo valley where
Pithecanthropus had been found. But shortly before its
departure in 1904 Selenka died. His wife Margarete Lenore,
who knew what was at stake, assumed leadership of the
expedition, assisted by the German geologist Elbert and
also by the Dutch mining engineer Oppenoorth – who was
later, in 1931, to discover the skulls of ancient Javanese
descendants of Pithecanthropus. The Selenka expedition did
indeed reap a rich harvest of animal bones, in exactly the
same state of fossilization as Pithecanthropus, but it found
no human remains. Nevertheless Frau Selenka was able to
demonstrate that all these fossils came from the same period
as the skull-cap and thigh-bone of Pithecanthropus, which
must therefore belong to one and the same creature. Today
the early men of Java are estimated to be about 450,000
years old.

Meanwhile the visitors' interest in Pithecanthropus had
awakened that of the local inhabitants. Shrewd Javanese
made the most of it, producing all sorts of bones of artificial
'ape-men', and selling allegedly prehistoric finds to the many
visitors at Trinil. A victim of such a forgery was a certain
Dr C. E. J. Häberlein. In 1926 he announced that he had

discovered a new Pithecanthropus skull, filled with a hard volcanic tufa. This announcement was taken up by the world press. But a mere glance at the photographs circulated by Häberlein was enough to convince experts that it was a single ball-and-socket joint from the upper arm of an extinct stegodon-elephant.

A participant at the Berlin and Leiden debates, the British anthropologist Elliot Smith, commented in 1931 about Dubois' finds: 'The amazing thing had happened. Dubois had actually found the fossil his scientific imagination had visualized! It is a marvellous chance that in the vast territories of the Sunda Islands he should have discovered this solitary fragment of skull, when forty years of intensive search since has failed to reveal another.'

But in the same year, that is, 1931, the study of Javanese man took a decisive step forward.

The head-hunters of Ngandong came to light. Ever since 1930 members of the Geological Survey of Java, among them ter Haar, von Koenigswald and Oppenoorth, had been busy comparing fossils from various sites in Java in an attempt to fit them into an over-all scheme of things so that they might reconstruct the prehistory of Java from the Tertiary to the Middle Pleistocene. In the course of this tremendous undertaking ter Haar discovered the small town of Ngandong, on the Solo river. Von Koenigswald describes this historic moment as follows: 'Ter Haar was an amiable colleague, tall, thin and always good-humoured despite his seven children . . . One day he came home rather earlier than usual, refreshed himself with a bath, and sat down to watch the sunset. As a geologist he was accustomed to take note of stones, and he suddenly observed a layer of gravel and sand about sixty feet above the present bed of the Solo, which must have been deposited by the river when it had flowed at that level . . . He could make out three different sets of terraces – the first about six feet above the river-bed, the second about twenty feet, and the third almost sixty feet.'

Opposite: Java man (Pithecanthropus erectus), a hypothetical reconstruction painting

173

It was an ideal spot for excavation, bound to contain fossils from all sorts of stages in Javanese prehistory. The Survey staff went zealously to work, discovering innumerable animal bones and naturally hoping to find fresh evidence of man. And soon afterwards they did: Oppenoorth dug up a number of skull fragments – by 1933 there were no less than eleven – which at first caused great excitement. Yet when von Koenigswald saw the skulls and photographed them he had to admit that unfortunately here was no new Pithecanthropus. The skulls were too lofty and too strongly vaulted, and the supra-orbital ridge – though massive and well developed – projected less sharply from the cranium.

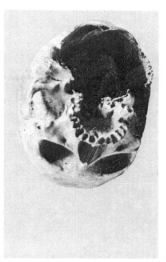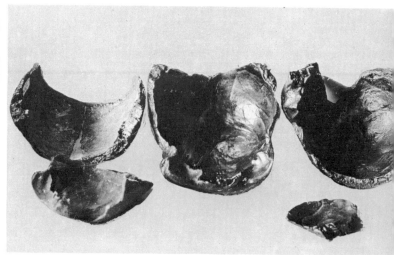

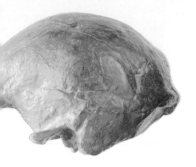

In spite of its new name, Solo man proved to be a Javanese Neandertaler.

The time factor seemed to point to this as well. Solo man, to go by the deposit in which he was discovered, must have lived at roughly the same time as the pre-Neandertalers of Europe. Nevertheless, there were some things which could not be accounted for. It soon became evident that Solo skull no. 1 had a brain capacity of only about 1000 c.c., which is much closer to that of Pithecanthropus than to Neandertal man. Skull no. 5 was also very different from the Neandertal type. At 22 cm. it was one of the longest if not the very longest yet found of fossil man. Fractures revealed that all these skulls had been violently smashed: hence Solo man must have been a head-hunter and cannibal. It is also possible that these finds are evidence of a prehistoric skull cult, as von Koenigswald believes: 'Perhaps the skulls were put down to mark off a particular area,' he writes. 'It seems that even today various tribes in New Guinea demarcate their dwelling or hunting grounds in a similar manner. They evidently suppose that the spirit dwelling in the skull can help them to defend an area against invaders.'

However the theory of a 'tropical Neandertal man' is now generally rejected. The anthropologist Franz Weidenreich was the first to point out that Solo man must be a direct descendant of Pithecanthropus. But it is ter Haar we have to thank for delivering Pithecanthropus from his previous state of isolation. His discovery of Ngandong was a real stroke of luck.

Skull fragments of Solo man found at Ngandong. The recent skull (bottom left) is a head trophy of the Dyaks. The occipital foramen has been smashed to facilitate extraction of the brain

A far older layer of earth was meanwhile engaging the attention of a certain Herr Cosijn, director of an iodine factory, and a keen amateur geologist. He had collected all sorts of remains of mammals from the hilly country near Modjokerto in eastern Java, and drew the Geological Survey's attention to this productive stretch of land. Soon a Malayan assistant, Andojo, while digging only about three feet down came across a strange small skull, the bones of which were as thin as the shell of an ostrich's egg. He sent it to the scientists with the comment that he had found the skull of an orang-utan.

In fact it proved to be the tiny skull of a Pithecanthropus child, about two years old. But when von Koenigswald and his Dutch colleagues announced this fact to the world there was an immediate protest from Dubois, whether out of envy, sheer scientific obstinacy or just old age cannot be

The skull of a roughly two-year-old Pithecanthropus child, found near Modjokerto in east Java

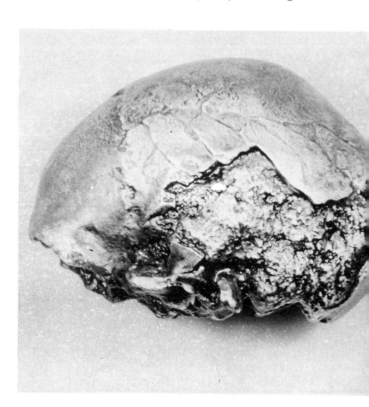

established. There was only one way of classifying the Modjokerto child with scientific accuracy, and that was to find another such skull, this time of an adult. The President of the Carnegie Institution, John C. Merriam, intervened at this point and provided the necessary funds for undertaking a systematic excavation of the similarly productive hills around Sangiran.

'The Javanese is very superstitious and dislikes coming into contact with human bones. Everything originating from the body can all too easily be misused for purposes of witchcraft. If he has a tooth extracted he takes it away with him and buries it where he is sure no one will find it. He is also extremely careful with hair and nails, and even with worn clothes . . . so that they shall not fall into the hands of anyone who might possibly use them to bewitch him.'

Thus to spur on local collectors von Koenigswald showed them fragments of a human skull and promised ten cents for every piece they brought him. 'But the result was depressing.

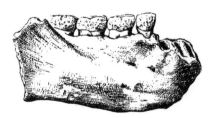

177

I had underestimated the business acumen of my brown collectors. They were breaking up the larger pieces behind my
back in order to sell me more.' Nevertheless three such
fragments were easily put together to make a complete
skull-cap. And near Sangiran von Koenigswald discovered
a fragment of a strikingly large chinless lower jaw, with very
odd teeth: the molars became larger towards the back of the
mouth, a characteristic only of anthropoid apes.

The reconstructed skull-caps turned out to be very interesting. When von Koenigswald compared them with
casts of Dubois' finds he was struck by a very strange thing:
'Without a considerable search it would be very difficult to

*Hypothetical reconstruction painting
of Pithecanthropus beside the Solo
River in Java*

find two recent skulls so similar as the skulls of Trinil and Sangiran, apart from the fact that the latter is more complete. Perhaps we need not be surprised that two individuals living at the same place at the same time should have resembled one another so closely. Nevertheless it was almost uncanny that two skulls should be found after such a tremendously long time which – in Weidenreich's words – were as alike as a couple of eggs.'

Further Pithecanthropus finds came to light, removing any doubts that still remained. Primitive Javanese man, so long the subject of conflicting opinions, is indeed a genuine early form of man, though in looks still very ape-like, and with a brain capacity of a mere 775–900 c.c.

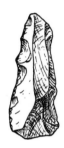

Stone implements from the Trinil stratum at Sangiran

His stone implements, too, were found in the deposit at Sangiran: 'They are small and obviously consist of flakes. As is usual with such pieces, one side is smooth and shows only the bulbar, or 'conchoid' cone displaying the concentric rings laid bare by percussion. The other, more irregular side has been retouched, or at least shows signs of wear . . . They are practically all scrapers. But there are also a few simple points, and one or two specimens have been so finely worked that they are entitled to be called primitive borers. The material is silicified limestone, yellow or brown, with a fine, lustrous patina which bears witness to its great age.'

But when von Koenigswald saw these implements he was already well acquainted with the tools and human remains lately found in a very different area of eastern Asia. This other discovery, the discovery of Peking man, was finally to convince even churchmen-critics of the correctness of the theory of descent. Strongly linked with this discovery is the remarkable Jesuit priest and palaeontologist, who also instigated a theological revolution, Pierre Teilhard de Chardin.

'God wills it so.' The town of Clermont-Ferrand, where this call went up at the time of the first crusade, is situated near the small country manor of Sarcenat, where, on 1 May, 1881, Marie-Joseph Pierre Teilhard de Chardin was born. His family was deeply religious, and at eighteen years of age he became a novice at the Jesuit academy of Aix-en-Provence. But, having since childhood had a passion for collecting stones, shells and fossils, he trained as a palaeontologist alongside his religious studies.

After his ordination he repeatedly took part in analyses of important prehistoric finds. He visited Piltdown, though without any suspicion of the fraud. He studied under the prominent French palaeontologist and anthropologist, Pierre Marcellin Boule, and accompanied Henri Breuil and Hugo Obermaier on their visit to Altamira. He dug up many fossils in many parts of the world, and is known particularly for his discovery of an early Tertiary primate, whose scientific name perpetuates his own. But best of all he liked to explore eastern Asia. In 1923 he took part in a scientific expedition to that part of the world. Three years later, with the support of the Rockefeller Foundation, he helped found an institute in Peking intended mainly to study the problem of fossil man. Members of six nations took part in this enterprise, among them the Canadian Davidson Black, the Swede Birgir Bohlin, the Chinese experts C.C. Young and Pei Weng Chung, and, briefly, Henri Breuil. Later one of the greatest American fossil hunters, Roy Chapman Andrews, who achieved world fame by his excavation of dinosaurs in the Gobi Desert, joined their ranks.

It is surprising to find a Catholic theologian of some eminence so constantly concerned to prove the existence of the missing link as prophesied by Darwin and Haeckel. But as one of his friends, Helmut de Terra, points out: 'The

claim that man is the most important being in the entire picture of evolution was not shattered, in Teilhard's opinion, by the anatomical considerations bound up with the early stages of human development.' In other words, Teilhard believed that the entire evolution of life – seen in an historical earthly perspective – only makes sense when man is considered the highest and unique manifestation of this process of evolution. Which makes us wonder why the quarrelsome churchmen who had criticized Darwinism had not hit upon this idea a little earlier.

Teilhard found the monogenetic descent of man from a pair of beings, as is described in Genesis, a story which was still firmly believed by most theologians of the time, to be unacceptable. He saw little hope, however, of proving it scientifically. We should be content to believe that man is as it were 'the axis and arrow of evolution', a 'unique mutation', striving after the greater development of his own consciousness. Thus Teilhard sees man to a certain extent as the product of a 'quantum transition' of the forces of life within the group in which zoologists have placed him, the Primates.

It was 'dragons' teeth' and 'dragons' bones' which put palaeontologists on the track of Peking man. Fossils had been used in China as medicinal aids from very early times. An instruction given to chemists at the time of the Emperor Chien-Lung (1771–99) runs: 'Dragons' bones are effective against heart, kidney, intestinal and liver ailments. They improve vitality, and have an astringent effect. They have a particularly beneficent influence upon the kidneys. As for nervous complaints, this medicine is particularly good for those in an advanced state of fearfulness, or for persons suffering heart attacks. Dragons' bones are also fine remedies for constipation, dreams, epileptic fits, fever, dysentery, consumption and haemorrhoids. Urinary diseases, as well as breathing troubles and boils can be healed by taking this medicine. Dragons' bones are equally effective as both astringents and laxatives. The best quality physic can be

The interior of a Chinese apothecary's shop (model of about 1880), where teeth of early man and giant apes were to be obtained

identified by the fact that when wrapped in silk it adheres to the tongue.'

Modern prehistorians are not inclined to ridicule this practice. Robert Broom, later one of the discoverers of a South African hominid, wrote: 'Some Western highbrows are inclined to sneer at the Chinaman's drugs, and say, "What virtue can there be in ground-up fossil teeth?" The Chinaman might reply that at least it never does any harm; which is more than can be said of some of the drugs in the British Pharmacopoeia.' In China there are families which have made their living for centuries out of digging up fossil teeth to trade as drugs. Von Koenigswald holds a similar view: 'These are only odd in our eyes. They are probably cheaper than many European medicines and just as effective!' It was precisely from these chemists' shops that he obtained some particularly interesting and instructive fossils.

Between 1899 and 1902 a German Dr Albert Haberer had collected a variety of fossilized mammal remains from Chinese apothecaries and the drug wholesale trade. Back in Munich his collection was studied and scientifically classified by the German palaeontologist Max Schlosser. As a result of Schlosser's report the Far East became a favourite stamping ground for prehistorians of very many nationalities. In fact it was a very strange molar in Dr Haberer's collection which put people on the track of a Chinese missing link. It could have belonged to an ape, but just as easily to an early form of man. Thus, in 1914, the first 'missing-link expedition', as Roy Chapman Andrews later called it, set out. Its most notable members were the Swedish geologist J.G. Andersson, the Austrian Otto Zdansky, and the fossil collector Walter Granger.

The difficulties which the expedition encountered are described by Roy Chapman Andrews, who later joined the team:

'One of the most serious was the universal belief in *feng-shui*, the spirits of earth, wind and water which guard all cemeteries. In the most thickly settled regions of China there are so many burial places that it is difficult to find a scrap of earth where *feng-shui* is inoperative. The fossil hunter must be extremely cautious in digging without having first obtained the consent of the nearest villagers. He needs unlimited patience, great tact and a saving sense of humour.

'Dr Andersson had some amusing experiences during his first investigations. Once when he had gone through all the necessary formalities of obtaining the owner's permission to excavate a fossil deposit, his operations were halted by the sudden appearance of an irate old lady. Angry men are bad enough, heaven knows, but when a Chinese woman works herself into a frenzy everyone hunts cover. This particular old lady was so enraged that she seated herself squarely in the hole the palaeontologist had dug and refused to move. Dr Andersson could not very well shovel her out except at the risk of having his face scratched; so, being a tactful gentleman, he tried making her ridiculous. It was a hot day and he borrowed an umbrella and gallantly held it over her head while the onlookers hugely enjoyed the performance. But the old lady only settled herself more determinedly and screamed even louder. Then Dr Andersson bethought himself of his camera, an instrument guaranteed to make any Chinese woman step lively. He politely explained to the spectators that without any doubt the old lady would like to have her picture taken while she was sitting in the hole. That was too much! Before the camera could be focused, she leaped out, screaming with rage. But even though she had been routed from her strategic position, she continued to create such a disturbance that Andersson's native assistants advised him to retire, leaving the enemy in possession of the field until the smoke of battle had lifted.'

Then in 1921 the evidence mounted. In a cave blocked by a fall of earth near Chou-kou-tien, Andersson and his colleagues found alongside the bones of animals some splinters of quartz obviously worked by hand. Two years later they found two ancient-looking teeth resembling the molar of Dr Haberer's collection which Schlosser had described. In an inspired moment Andersson risked a dramatic prophecy: 'In this spot lies primitive man. All we have to do is find him.'

184

These discoveries made a deep impression upon the Swedish Crown Prince Adolf, now King Gustav VI, who took a keen interest in scientific matters. So, with Swedish assistance and that of the Rockefeller Foundation, a second 'missing link expedition' was launched, that exemplary excavation which lasted from 1927 to 1939, and which had among its leading participants Davidson Black, Teilhard de Chardin, Roy Chapman Andrews and Franz Weidenreich.

Between 16 April and 18 October, 1927 alone the expedition sifted 3,000 cubic metres of earth on the 'Dragons' Mountain' of Chou-kou-tien, under the direction of the Swede, Birgir Bohlin. A tremendous number of fossils came to light, but only one single human-looking tooth. 'It was a remarkable tooth', writes von Koenigswald. 'Black had never seen anything like it. He made a detailed comparison of this tooth with human and simian teeth and came to the conclusion that it was a relic of a hitherto unknown fossil man, whom he designated *Sinanthropus pekinensis*. It was certainly audacious to label a new prehistoric man on the basis of this one tooth. But Black felt justified in doing so.'

The human-looking molar dug up from the 'Dragons' Mountain at Chou-kou-tien by Birgir Bohlin in October 1927

185

In the next two years the team dug up a further 9,000 cubic metres of earth and sent about 1,500 boxes of fossils to Peking. But of Peking man himself they found only a few odd teeth and some fragments of a lower jaw. Then, on 2 December, 1929, in a bitterly cold winter, the labours of these by now disheartened men were finally rewarded.

'Roy, we've got a skull!' exclaimed Black to his colleague Andrews. The Chinese leader of the excavation, Pei Weng Chung, had come across the skull in a bank of hard limestone. It took four months to free it of the surrounding stone. It was relatively small and low, as von Koenigswald reports: 'It had virtually no forehead and a continuous, bony supra-orbital ridge, such as we only see among anthropoid apes, projected sharply over the eyes. None the less, anatomically there could be no doubt that it belonged to a primitive man. Not only was its cranial capacity of about 1,000 c.c. too large for an ape, but also the brain-cast clearly showed an approach to human proportions. Completely

Excavation work in progress at Chou-kou-tien

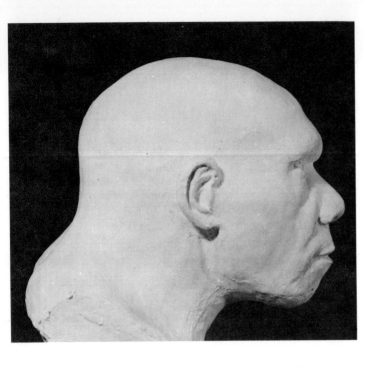

Reconstruction bust of Peking man

human, too, was the fossa of the temporo-maxillary joint, which was deep-cut as in modern man, not shallow as in apes.'

Representatives of eight nations took part in the party organized to celebrate this find, among them Teilhard de Chardin, who considered the discovery of Peking man to be of really decisive importance. He was not only fascinated by the skull but also by the stone implements Sinanthropus had used: 'What happened between the last strata of the Pliocene age, in which man is absent, and the next, in which the geologist is dumbfounded to find the first chipped flints? And what is the true measure of this leap? This is the question we must answer before we follow the march of mankind step by step up to the decisive state of transition through which it must go.'

This was probably the first time a Catholic theologian had expressed his belief in the general, progressive evolution of man. Roy Chapman Andrews, a sober man of facts, saw things more simply when he viewed the skull of Peking

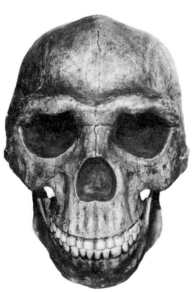

Peking man already knew the use of fire: hypothetical reconstruction painting by Maurice Wilson. Below: reconstruction model of the skull of Peking man

man: 'There it was, the skull of an individual who had lived half a million years ago, one of the most important discoveries in the whole history of human evolution! He couldn't have been very impressive when he was alive, but dead and fossilized he was awe-inspiring.'

'Little white musk-rat' was the name the Indians gave Davidson Black back home in Canada, where as a young man he had accompanied them on a number of canoe trips. He was slim, thin-faced, and not very strong. In fact he should have left the Far East long ago, for the climate did not agree with him, but such was his enthusiasm for Peking man that he kept on postponing his departure. He is once reported to have exclaimed to a visitor: 'Why, man, just think of being paid for doing the one job in the whole world you've always wanted to!' But on the morning of 15 March, 1934, his secretary found him slumped over his desk, the skull of Peking man in his hand. He had suffered a fatal heart-attack.

Meanwhile excavations continued at Chou-kou-tien, and more skulls and bones of Peking man were brought to light.

from Java von Koenigswald hurried to the spot, compared his Pithecanthropus with the Chinese Sinanthropus, and agreed with his colleagues that the two forms were related or perhaps even the same. Both had lived at roughly the same time, at the most half a million years ago (on this point scientists do not entirely agree). The differences between the primitive Javan and Peking types were mainly cultural. Peking man had obviously been a social creature, and was already acquainted with the use of fire. Von Koenigswald reports: 'At the beginning the site of the find had resembled a cave. It now proved to be a wide cleft, in which prehistoric man had sought shelter . . . Stratum lay upon stratum, many of them full of bones and ashes, stone implements and jaws . . . a vast accumulation of bones showed that Peking man had been a skilful hunter. The quantity of his refuse seems to have attracted a large number of hyenas . . . Altogether the scanty remains of some forty-five Peking men were brought to light . . . All the skull parts are known with the exception of one – the skull base. For Peking man was found to have possessed a very human trait: he was a cannibal. All the skulls were smashed, and the region around the foramen magnum was broken away in order to extract the brain, and there was even a femur that had been split lengthwise to get at the marrow.'

This encounter with Peking man was a fateful one for Teilhard de Chardin. Even before he had left Peking he had begun work on his book *The Phenomenon of Man* – a vision of the origin and general destiny of man from the viewpoint of a modern theologian drawing on his profound knowledge of the natural sciences. Although Teilhard de Chardin's ideas and understanding first met with opposition from his Church (the Vatican even placed his writings on the Index), the time was ripe for views such as his, and for Christianity and the theory of descent to patch up their quarrel. He wrote to his friend Helmut de Terra: 'I found that far more is thought of my moral standpoint than I had expected. I would like to believe that the present events in society have opened men's eyes to the true nature of these processes. My

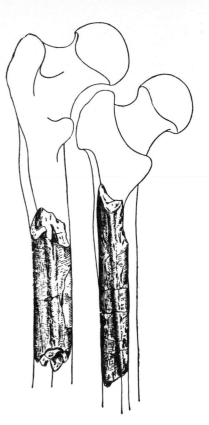

Thigh bones of Peking man. Below: our ancestors in China already used bone and antler to make tools

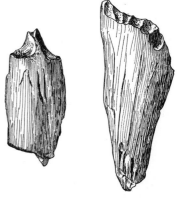

189

AFRICA ASIA	EUROPE		AFRICA	ASIA	EU
Late Pleistocene	Upper Pleis-tocene	Würm Ice Age	H. erectus rhodesiensis	H. erectus soloensis	
		Riss-Würm Inter-glacial	H. erectus rhodesiensis		
Middle Pleistocene	Middle Pleis-tocene	Main Inter-glacial	H. erectus mauritanicus	H. erectus mapaensis H. erectus pekinensis H. erectus erectus	
Early Middle Pleistocene	Lower Pleistocene		H. erectus capensis? H. erectus leakeyi		H. erec heidelb ensi

Finds relating to early man (Homo erectus), arranged regionally by eras

various writings have been widely read, and do not seem to have offended anyone.'

Teilhard de Chardin already thought that probably neither Java nor Peking man should be considered the direct ancestor of man. He considered the primitive Asians to be 'verticils', as it were, on the family tree of man, forms whose change had been so rapid that they soon died out, while the real pre-sapiens type developed alongside them, growing so dominant as to become the sole surviving form.

The accepted view now is that the pre-sapiens race had come into existence while Asiatic early man was still around, and that the last traces of early man lived on to overlap with the developed culture of the Late Ice Age. Whereas some anthropologists still like to think of Java man as the ancestor at least of the aboriginal Australians and Peking man of the Mongoloid races, the majority now envisage something rather more complicated, a sort of 'genealogical bush', with many offshoots.

Then came Pearl Harbour, with a sudden and disastrous effect upon the further study of Peking and Java man. Von Koenigswald was interned on Java by the Japanese. He writes: 'All the finds so far described had to be placed under their supervision in the Geological Survey's big safe. We very circumspectly substituted plaster casts for some of the originals. The casts were extremely well made and to lay eyes almost indistinguishable from the originals. We had mixed finely ground brick dust with the plaster of Paris, so that even in the event of injury the break would remain nicely dark, as in a genuine fossil . . . The new finds, not yet described and therefore unknown, were housed with neutral friends – a Swiss geologist and a Swedish journalist. Showing great temerity, my wife retained the precious upper-jaw. All the finds survived the war. Our Swedish friend, fearing a house search one day, put my collection of teeth – which included those of Pithecanthropus and Gigantopithecus – in large milk bottles and buried them by night in his garden.'

Flake-tool of chert, for scraping, found at Chou-kou-tien

Opposite: much older than Peking man, so long reckoned the true Adam, is the African Australopithecus, the discovery and study of which is described in Part IV

Thanks to these precautions all the Java finds survived the war. But the fate of Peking man was a tragic one. One of the most beautiful and best preserved skulls unearthed on the Dragons' Mountain of Chou-kou-tien had been nicknamed 'Nelly' by the anthropologists; this Nelly, in von Koenigswald's words, 'looked out into the world from beneath a thick supra-orbital ridge, fresh, intelligent and unperturbed, as only a person can look who has never been to school.'

'Where is Nelly?' A postcard bearing this inscription arrived in New York, and American scientists knew immediately that something had happened to the Peking finds. An attempt had been made shortly before war broke out to ship the valuable fossils to the United States. But by the time the American transport ship arrived at the harbour to meet the train carrying the boxes from the Peking Institute, the bombs of Pearl Harbour had fallen; since then there has been no trace of the skulls and bones from the Dragons' Mountain. After the war all sorts of rumours circulated. According to one, the barge taking the boxes to the freighter capsized; another would have it that the Japanese who captured the train simply threw them away as worthless; a third that the theft was perpetrated by Chinese dockworkers who then sold the dragons' bones to local chemists.

Be that as it may, we are indebted to Chinese palaeontologists and anthropologists not only for the fact that we have good casts and excellent descriptions of the lost fragments, but also for a number of more recent finds made in different regions throughout eastern Asia.

Chinese pride in their ancestry has led many anthropologists from Peking man's homeland to assume, despite assertions to the contrary, that these discoveries at Chou-kou-tien represent no less than the ancestors of the Chinese people of today. Thus the discoverer of Peking man, Pei Weng Chung, together with his colleague Wu Yu-kang and other anthropologists, began to look for further prehistoric evidence in various provinces of the Chinese Peoples'

Republic. In 1963 a well preserved lower jaw was found in Lantian, in the province of Shensi. A year later a group of field workers of the Peking Academy Institute excavated a skull-cap, rather more primitive in appearance than that of Peking man, about twenty miles from there. More fragments of bones, teeth and stone implements, similar to those found at Chou-kou-tien, were discovered by the Chinese in the North Vietnamese province of Lang Son, mainly in the 'Tiger Cave' near Luc-Yen. And more finds were made in Java. Obviously a very wide variety of human types had lived in eastern and southern Asia at various times. The most advanced groups headed north, into the Asian continent and the hinterland of India, whence they peopled the Indonesian islands. But where had these early men come from? Did they originate in Asia itself?

There were giants once on earth, so the Bible tells us. Franz Weidenreich, one of the ablest men to study Peking man, recalled these Old Testament words when von Koenigswald showed him a very strange tooth, remarkable for its great size. This giant tooth had been discovered in a chemist's shop in 1935. Later von Koenigswald searched the

Opposite: Today it is no longer the excavations of Asiatic early man which dominate the study of man's evolution, but the discovery of the African australopithecines. Above: looking across the Olduvai Gorge in Tanzania where Mary and Louis Leakey made their great discoveries. Below: Find spot of Zinjanthropus, one of the australopithecines described in Part IV

Chou-kou-tien, August 1931: Pei and Wong, the Chinese discoverers of Peking man, with the Abbé Breuil

shops for more of them. He thought they were the teeth of some great ape, and endowed this unknown creature with the name *Gigantopithecus blacki*, 'Black's Giant Ape', in honour of his late friend Davidson Black. The molars were so large that some people believed von Koenigswald to be wrong. They thought they must be the giant molars of at least a boar or a bear.

And gradually even von Koenigswald began to doubt his own interpretation: 'They bear an astonishing resemblance to human teeth . . . An upper canine in my collection probably also belongs to this giant form. It has a very strong, completely straight root. In anthropoid apes the root of the canine is always markedly curved, because of the formation of the muzzle, and the crown tapers to a point. This canine would fit much better into a human face than an ape's muzzle. Even if Gigantopithecus were a giant man . . . he would still not have any direct connection with our family tree . . . In any case we must assume that Gigantopithecus belonged to a collateral line. Thus these teeth present us with problems to which no one has yet found an answer, and we can only hope that some geologist will eventually succeed in solving the mystery.'

Fragment of a lower jaw-bone of 'Meganthropus palaeojavanicus', presumably a relative of the African robust Australopithecus

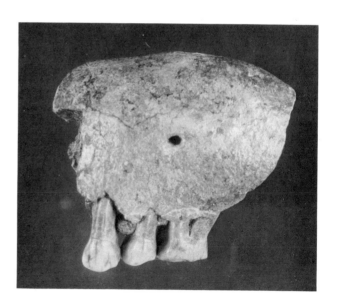

Another view (from above) of the lower jaw of the 'giant man' of Java

'One of the few general rules in palaeontology** is
that more primitive, earlier forms are generally smaller than
their specialized and larger descendants', von Koenigswald
once wrote. But neither the Chinese giant teeth nor some
new finds made in Java fitted this theory. In Java a particularly large Pithecanthropus skull had been found with a gap
in the lower jaw between the canines and the lateral incisors
– the so-called 'simian diastema', previously thought to be
a characteristic of apes but not of man. It was christened
Pithecanthropus robustus, and is today considered to belong to
the same group as the Modjokerto child.

But some bones which proved even more extraordinary
were the fragments of a powerful lower jaw which von
Koenigswald obtained from his collectors in the spring of
1941. It was as large as the jaw of a gorilla, and had three
colossal teeth; but despite this it looked remarkably human.
Another fragment of jaw, found earlier at Sangiran in 1939,
again looked rather different and was again of astounding

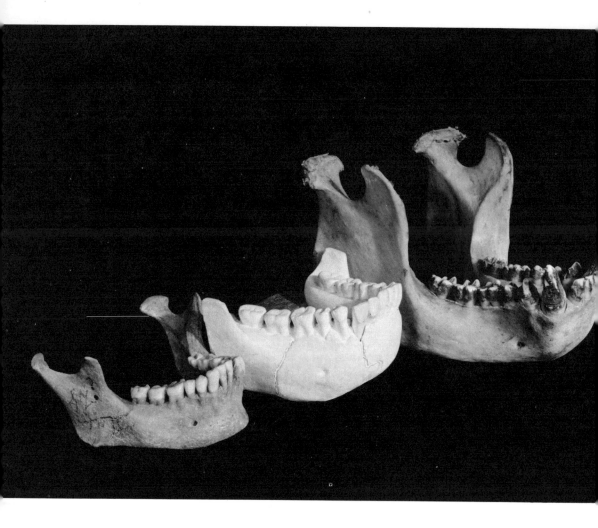

Comparison of the jaw-bones of modern man (left), Meganthropus (centre) and gorilla (right)

size. This puzzling creature was given the name *Meganthropus palaeojavanicus*, Great Man of Ancient Java. The deposits in which the bone fragments were found quite clearly showed the Javanese giants to be far older than the early men so far known – von Koenigswald estimated them to be more than half a million years old.

'I believe that all these forms have to be ranged in the human line and that the human line leads to giants, the farther back it is traced.' This was Weidenreich's amazing

196

conclusion as he studied the casts in New York. 'In other words, the giants may be directly ancestral to man.'

It was during the Second World War, when von Koenigswald was rumoured to have been drowned in Java, that Weidenreich felt he should write a description of these Asiatic giants. On 9 May, 1944 he gave a lecture to the Ethnological Society in New York which for many years after was remembered as a truly historic occasion. He led his hearers to believe that the Chinese Gigantopithecus should be considered the initial stage in the evolution of man. In China this giant form had given rise to Peking man, who was the direct forefather of the Mongoloid races. But in Java it first developed into Meganthropus, and then, like the rungs of a ladder, evolved into *Pithecanthropus robustus*, *Pithecanthropus erectus*, through Solo man and Wadjak man to the present-day Australian aboriginals and Papuans.

As Willy Ley, the American palaeontologist, put it, Weidenreich's lecture 'was received by the meeting with exclamations of incredulity which politeness could hardly disguise'. Nevertheless, Weidenreich expanded the same theme in his book, *Apes, Giants and Man*. With many types of animals, he explained, domestication of the species results in the production of a smaller animal. Such is the case, for instance, with ponies, dwarf donkeys, dwarf pigs, dwarf bulls, and miniature breeds of dogs. As the evolution of man is really a slow process of domestication, we must assume our primitive ancestors to have been giants. What is more, Weidenreich pointed out, dwarf forms usually have diminished jaws and higher foreheads – as is the case with modern man.

The Chinese giants are still a mystery today. Between 1956 and 1963 Chinese palaeontologists, under the leadership of Wu Yu-kang, discovered three almost complete lower jaws and more than a thousand odd teeth belonging to this gigantic creature. As Wu Yu-kang says, the lower jaws and molars are larger than those of any living or extinct form of man or ape; on the other hand the incisors and canines are smaller and less pointed than those of great apes living today. Because of the nature of this giant's molars Wu Yu-kang thinks he should be regarded as a hominid, a form preliminary to man, and to be regarded as similar to those African hominids who are discussed in the next section of this book.

But the Chinese giant jaw-bones are of the Pleistocene period and therefore more recent than most of the African hominids so far discovered. In the Siwalik hills below the Himalayas the American palaeontologist Elwyn L. Simons found a fourth Gigantopithecus lower jaw of the Pliocene period, which means that the 'giants' existed as far back as in the Tertiary, thus presenting us with another riddle.

In Wu Yu-kang's opinion the Chinese giant will have walked relatively upright, so that he was able to use his hands for holding tools and weapons. This is why his incisors and canines had become so small.

Most Western anthropologists believe that the Chinese giants are an extinct giant ape. But in Africa, on the other hand, as we shall soon learn, evidence has been found of a unique race of sturdy creatures which must unquestionably be classified as hominid. And the Javanese giants suggest that they too should be interpreted as such. Thus our present evidence makes it look as if Asia must surrender to another region of the globe her position as the place most likely to have brought forth man.

Now the trail leads us to Africa. It has been known since 1935 that not Eurasia alone was the habitat of early man, for in that year the German doctor and ethnologist Ludwig Kohl-Larsen collected a number of blackish-green metallic fragments of bone on the shores of Lake Eyasi in east Africa. Hans Weinert, one of the leading German anthropologists of the day, assembled the fragments into a skull-cap and announced that Kohl-Larsen had found an east African counterpart to Peking man.

Since 1954 a group of French prehistorians under the leadership of C. Arambourg have dug up a parietal bone, three lower jaws and many odd teeth from a sandpit near

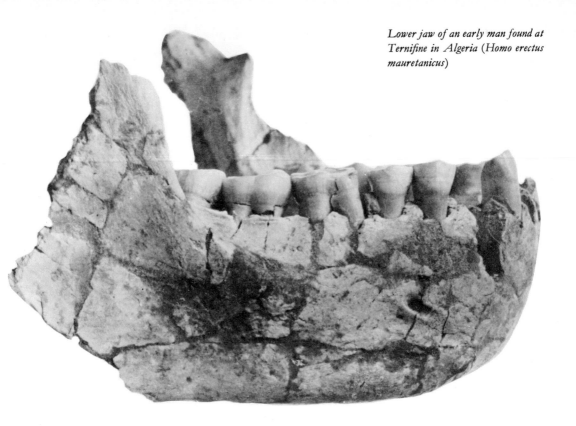

Lower jaw of an early man found at Ternifine in Algeria (Homo erectus mauretanicus)

Ternifine in Algeria. They were given the name of *Atlanthropus mauritanicus* and were likewise classified as early man. More remains of Atlanthropus were also found near Casablanca, in Morocco. And a human skull which that great discoverer of fossil man, L.S.B. Leakey, washed out of Bed II of the Olduvai Gorge seems also to belong to early man. Olduvai, in east Africa, is probably the most famous place for the discovery of successive periods in the evolution of mankind, and Bed II, the second stratum from the bottom, contains stone implements similar to those found with the Asiatic early men.

By the time Leakey made the discovery of the human skull at Olduvai the many confusing generic names of early men (Pithecanthropus, Sinanthropus, Palaeanthropus, Africanthropus, Atlanthropus) had been more or less abandoned.

Hand-axe made of petrified lava, found in Bed II at Olduvai

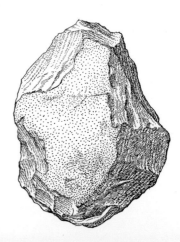

All these forms are now classified as belonging to the genus *Homo*, to which we too belong; but their classification as *Homo erectus* brings them together in another way. The Javanese, Chinese, European, north and east African early men probably all belonged to the one group, which spread gradually over the entire Old World and to which Heidelberg man will also have been related.

Thus the genus *Homo* can be divided into three groups, which to some extent comprise an evolutionary sequence:

Skull fragment of East African early man, Homo erectus leakeyi, from Bed II at Olduvai

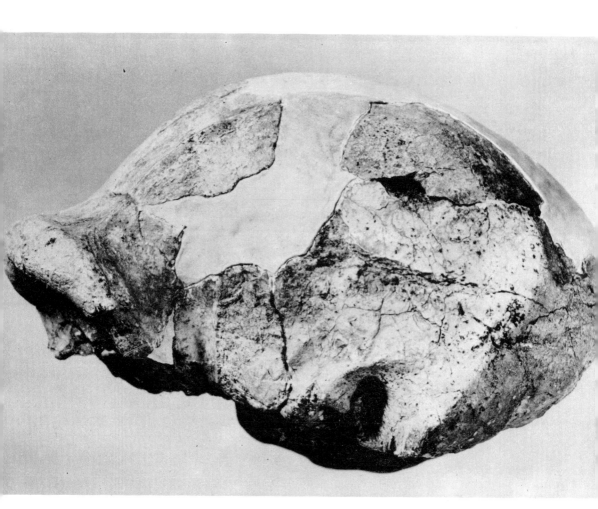

1. The Erectus group
2. The Neandertal group
3. The Sapiens group

The 'giants', however, do not fit into this rather restricted classification of types. It is only recently, over the last few decades, after leading investigators had discovered other similar giant forms in Africa as well, besides hunting and weapon-bearing pygmies, that the scientific world has begun to take them seriously. The Javanese meganthropus, at least, seems to be closely related to or even to some extent identical with the African 'Nutcracker' man, whom we shall later describe.

The trail which the eccentric Eugène Dubois had picked up in Java, which von Koenigswald had followed on the banks of the Solo River and in Chinese chemists' shops, and which an international research group including Davidson Black, Franz Weidenreich and Pei Weng Chung had then taken up in China, now leads us into the shrub-grown plains, chalk caves and rock crevices of Africa. Here, if we are to trust our present scientific knowledge, the real story of man's creation begins.

Part Four

Between animal and man

The most cowardly thing of all is to fear
that the truth could be bad.
HERBERT SPENCER

DART
BROOM
ROBINSON
BRAIN
LE GROS CLARK
RECK
LEAKEY
EVERNDEN
COLE
HEBERER
TOBIAS
MORRIS
VAN LAWICK-GOODALL
KORTLANDT
NAPIER
JULIAN HUXLEY

'The Missing Link is a catch phrase – a sure-fire leader for any newspaper in the world.' With these words Roy Chapman Andrews prefaced the announcement of the first of a series of shattering discoveries in Africa, discoveries which were to help fill the gaps in our picture of the origin of man, alter it and yet at the same time confirm its general outline in a surprising and indeed really magnificent manner. 'But not only *one* link is missing from the broken chain of human ancestry. Dozens remain undiscovered. Still, we have enough with which to reconstruct the chain pretty accurately. But the only Missing Link that interests the public is a sub-human, combining the characters of both ape and man. Just such a Missing Link has been discovered! ... It has always been obvious that when, or if, such a creature that stood at the fork where man took the high road and apes took the low road ever was found, it would make its debut unheralded and unsung. Its ape-man characters would be so obscure, so technical, that they could be discerned only by an anatomist. It never would fulfil the public's preconceived picture of the Missing Link. It would look so much like an ape and so little like a human that no newspaper reporter, no matter how imaginative, could stimulate popular appreciation.'

Even at the time of the great Gold Rush, towards the end of the last century, gold-diggers and playing children will probably have come across quite a few fossils of South African ape-men. For instance, a lawyer later recalled 'playing bowls with a stone skull as a child' near Sterkfontein, a site which has since become famous in the annals of prehistoric man. Yet, just as so many great discoveries and so many break-throughs in knowledge have at first gone practically unnoticed in the history of mankind, so the solving of the problem of our origin has gone similarly unheralded. The first discovery was made near Buxton, on the edge of the Kalahari, near a railway station then called Taung, and it caused no sensation at all. The world press as good as ignored it; nor was much notice taken of the discoverer's entirely accurate assessment of it. The few experts

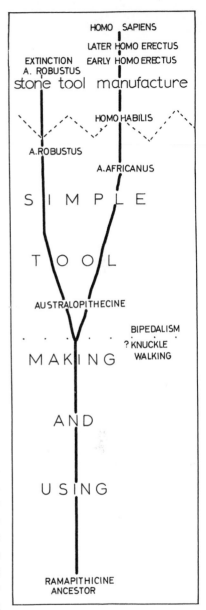

Possible interpretation of hominid evolution

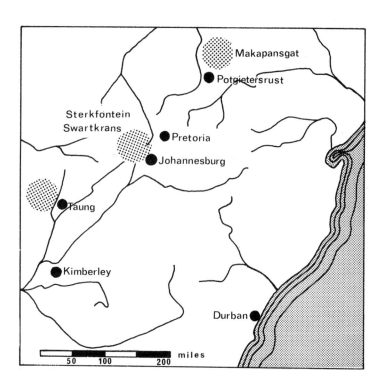

Map of the most important sites in South Africa where the remains of australopithecines have been found. They total more than a hundred

who did listen to his interpretation greeted it at best with scepticism, more often with disbelief, or even scorn. The man in question, Raymond A. Dart, who set in train the great wave of African discoveries, is vividly portrayed in *African Genesis* by his friend and biographer, Robert Ardrey:

'Dart is a small, compact man of far-reaching interests, far-gripping personal magnetism, and appalling durability. Until recent years he still gave lectures to his astounded class in comparative anatomy while brachiating cheerfully from the steam pipes over its pupils' heads. I recall an occasion a few years ago when the two of us were climbing a steep wall of the wild Makapan valley ... to visit an unhappily situated cave. Half-way up my breath went out of me as from a punctured tyre. We stopped. "Yes", said Dart, gently, compassionately, breathing as easily as a sleeping

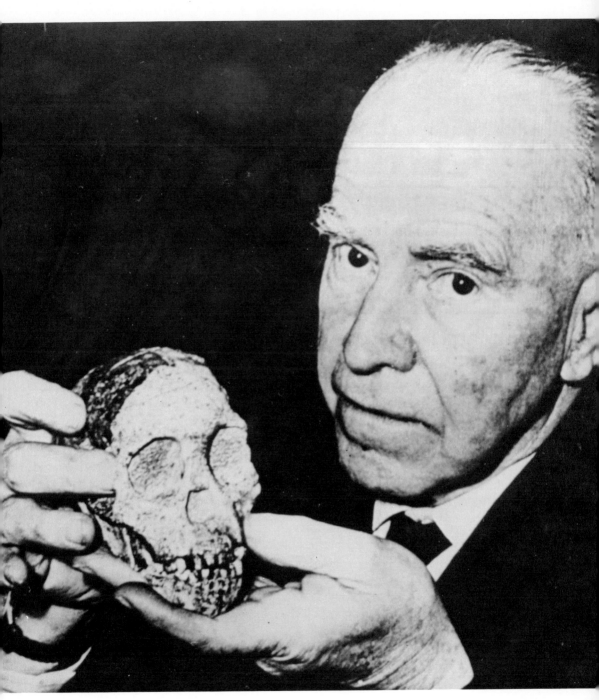

child, "it's a difficult climb." I reflected without pleasure that Dart was all of sixty-five years old. And he was smiling to himself in pleasant reminiscence. "Do you know," he said, looking about as if he had just discovered something, "this is exactly the place where old Broom always had to stop." I reflected with even less pleasure that Robert Broom . . . had not even entered the field of anthropology until he was seventy.'

This remarkable man was born in Australia, studied in England and the United States, and in 1922 went to South Africa as an anatomist. Only two years after his arrival he came across a very special fossil, the child of Taung, which was later to make him a famous and controversial figure. His own report of this portentous discovery runs:

'The South African "missing link" story goes back to 1924 when the late Miss Josephine Salmons, then a young science student in anatomy, brought me a fossil baboon skull that she had found on the mantelpiece of a friend she had visited the previous Sunday evening. It had come from the Northern Lime Company's works at Buxton, and was the first intimation that any fossil primate had been found in Africa south of Egypt. So we became very excited, and after interviewing the professor of geology, Dr R.B. Young, learned to our satisfaction that he was going to Buxton the following week.

'Arriving at Buxton, Professor Young learned that in the previous week a miner, M. de Bruyn, had brought in a number of fossil-laden rocks blasted out the week before. When they came to Johannesburg I found the virtually complete cast of the interior of a skull among them. This brain cast was as big as that of a large gorilla: and fortunately it fitted at the front end to another rock, from which in due course there emerged the complete facial skeleton of an infant only about five or six years old, which looked amazingly human. It was the first time that anyone had been privileged to see the complete face and to reconstruct accurately the entire head of one of man's extinct ape-like

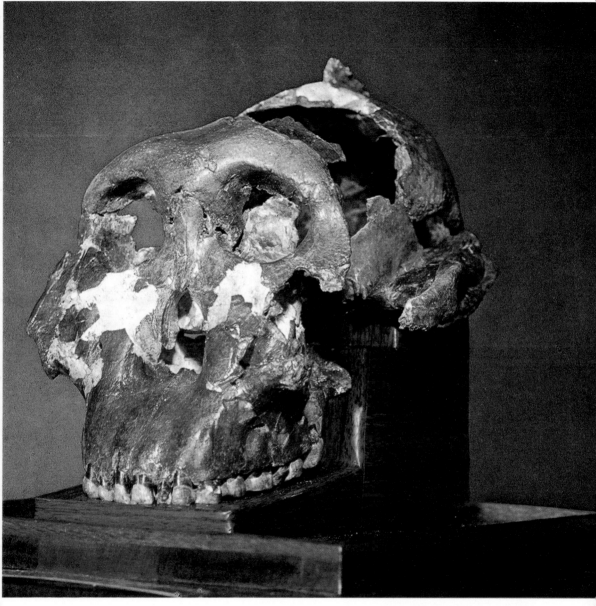

The famous cranium of 'Nutcracker man' from Olduvai. Later designated Zinjanthropus boisei, it is today referred to as the robust Australopithecus boisei

Louis Leakey and his wife Mary compare the skull of a South African australopithecine with Zinjanthropus which they found at Olduvai

relatives. The brain was so large and the face was so human that I was confident that here indeed was one of our early progenitors that had lived on the African continent.'

Dart was a cautious man. First he thought the Taung child was merely a very human-looking young ape, and therefore gave it the scientific name *Australopithecus africanus* (Southern Ape of Africa). But then on 7 February, 1925 he published a prophetic statement: 'The specimen is of importance because it exhibits an extinct race of apes intermediate between living anthropoids and man.' This was enough to turn the entire body of established anthropologists against him. It had taken them long enough to assimilate the idea that man had originated in East Asia, and now here was someone declaring this was not the case at all! The 'Taung baby' was estimated to have lived about a million years ago, but the imperfect dating methods of the day reckoned the Asiatic early men to be just as old, if not older. As always, Dart's conjectures did not fit in with the prevailing theories.

The experts were highly critical, and this ensured that the 'Taung baby' played no part in any scientific discussion for another ten years or more. Sir Arthur Keith, the doyen of British anthropology, declared with the full force of his authority: 'It is a true long-headed, or dolicephalic, anthropoid – the first so far known.' Sir Arthur Smith-Woodward, although so completely taken in by the Piltdown forgery, was deeply suspicious: 'There are serious doubts in the material for discussion, and before the published first impressions can be confirmed more examples of the skull are needed.'

The reactions of leading German anthropologists were just as negative. 'In any case it is an anthropoid ape, not a hominid,' declared the by now venerable Hans Weinert. His colleague, Wilhelm Gieseler, completely rejected Dart's hypothesis: 'Australopithecus is not a man, nor does he belong directly to our gallery of human ancestors.' Palaeontologist Wolfgang Abel was more explicit, declaring that the Taung child was closely related to the gorilla, but had 'never attained the level of specialization of the modern gorilla, which is very high in many ways.'

When Dart countered by saying that one should first see the skull before passing such judgments upon it, the Taung fossil was dubbed 'Dart's child'. Most people reckoned it to be a fossil chimpanzee. It was soon forgotten, and its

The Taung skull, side view

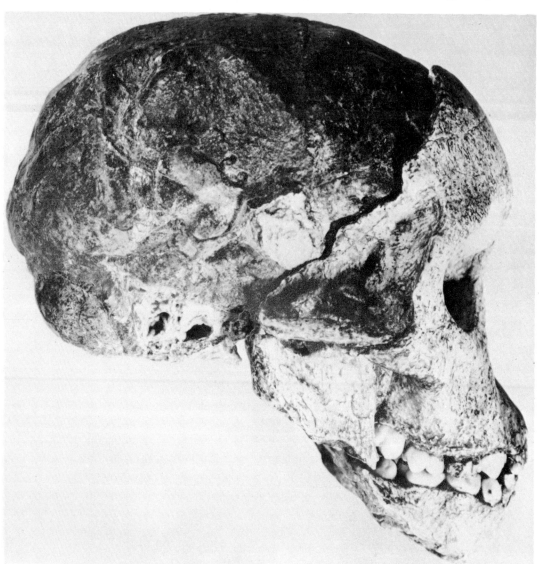

discoverer dismissed as a scurrilous outsider bent on making a name for himself by producing a new and completely misleading theory. It was only when another man came upon the scene, one who had already made a name for himself by his discovery of the missing links between reptiles and mammals, that the case was taken up again.

The Taung skull, frontal view

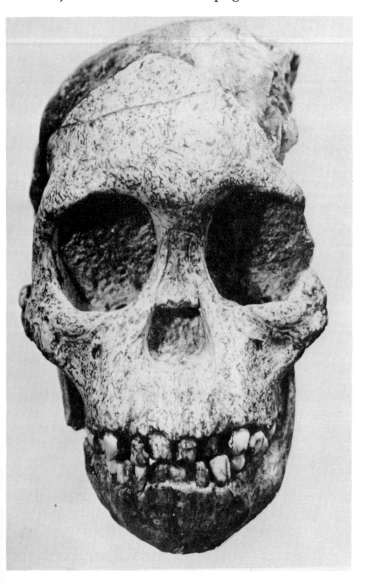

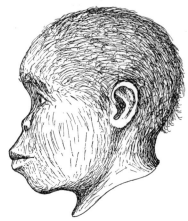

Robert Broom's reconstruction of the head of the Australopithecus child found by Dart

In order to recuperate from an attack of bronchitis
Robert Broom was, at the age of six, sent to the coast for a
time to live with his grandmother. Even in these childhood
years he developed a passion for collecting natural objects.
Son of a Scottish textile designer, he later studied anatomy,
became a doctor, and when his studies had ended went to
Australia, intent on using his spare time there to study the
country's ancient marsupials whose ancestral past was as yet

Robert Broom, engaged in excava-
tion work in South Africa

unknown. In 1896, when he was thirty, he heard that fossilized remains of mammal-like animals had been discovered in the Cape, so he resettled in South Africa and began his search for the transitional forms between reptiles and mammals which had not yet been discovered and were of particular importance for the study of the origins of species.

Like Dart, Broom conducted his researches with great discretion, *in camera*, as it were. He took up general medical practice on the borders of the Karroo, a barren area renowned for its wealth of fossils, and devoted every spare minute of his time to excavation. In fact he succeeded in discovering so many transitional forms between reptiles and mammals that in 1906 he attracted the notice of the great American palaeontologist, Henry Fairfield Osborn, who encouraged him to continue his good work.

In order to devote himself more intensively to his beloved fossils Broom gave up his medical practice, even though it had become well established, and took up an appointment as Professor of Geology and Zoology at Victoria College in Stellenbosch. Later Jan C. Smuts, Prime Minister of the former Union of South Africa, wrote the following tribute to him in a memorial volume published in Broom's honour.

'His palaeontological researches on the Karroo reptiles enabled him to bridge successfully the gap between the reptiles and mammals. This gap had long remained the most considerable in animal palaeontology ... In the Karroo reptilian fossils he found a rich horde of evidence which he carefully analysed and correctly interpreted in a series of researches as masterly as any made during this century. The story of the evolution of life on this globe is perhaps the most enthralling romance in all science. Much of it remains obscure or unknown, but much more was debatable or unknown when Broom embarked on his Karroo researches.'

Broom was the only person who as a result of his thorough study of the Taung skull came down immediately and unreservedly on the side of his younger colleague. Dart had been so hurt by the treatment he had received that he would undertake no further research, so Robert Broom, this

213

Hypothetical picture of 'Dart's child'

'grand old man of South African prehistory', plunged in where Dart had left off. His examination of the teeth of the Taung child convinced him that this controversial creature was more closely related to man than to any anthropoid ape. Also, the position of the brain led Broom to assume that it had already been able to walk upright. Nevertheless he realized that Dart's view could only be proved if further specimens of Australopithecus were discovered in the caves and quarries of South Africa. But, he declared, it was no use adopting a defensive attitude; the world must be told of the possibility of finding missing links in this region of the world so that when it happened it would be psychologically prepared.

A regular press campaign, inspired by Broom, drew the attention of quarry owners, students and tourists to the possibility of turning up fossils of true scientific value in South Africa. The small town of Sterkfontein even issued a brochure containing the now classic words: 'Come to Sterkfontein and find the Missing Link!' This was just Broom's way. Despite his age, he knew exactly how to go about a problem, doing so with verve and sparkling wit. When at

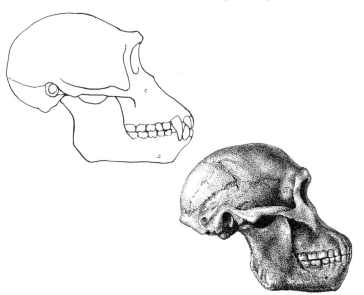

A chimpanzee skull compared with that of Australopithecus transvaalensis africanus

an earlier stage of his life the bunglings of bureaucracy had caused him such irritation that he had abandoned his palaeontological work and retired as a doctor to a provincial town, instead of entrenching himself behind a barricade of bitterness, he had in the space of three short months succeeded in becoming medical officer to two goldmines, president of both the chess club and the revolver club, and finally even mayor of the town. Robert Ardrey, who tells this story, describes Broom, perhaps the best known South African scientist before Christiaan N. Barnard, in these vivid words:

'He was a tiny man who dressed even in the midst of the bush in black hat, black tie, and stiff white collar; and I question that a more zestful figure has ever hustled through a Permian landscape. Broom was a doctor as well as a zoologist, and an accumulator besides. He collected everything: fossils, Rembrandt etchings, postage stamps, susceptible girls.'

In the end he collected fossil men too. In the summer of 1936, already seventy years of age, he arrived at Sterkfontein, the town which had challenged tourists to find the

Robinson and Broom working on fossils at Sterkfontein

missing link, and there proceeded to find it. G.W. Barlow, who supervised the local quarrying works and who was in charge of the caves, was shown the Taung skull and admitted that he often found things like that and sold them to the tourists as souvenirs. Broom made him promise to let him have the next piece which should come to light, in the cause of science, and this Barlow did. A few days later he handed this venerable palaeontologist a splendidly preserved skull of an adult australopithecine, though unfortunately lacking its jaw.

A schoolboy, Gert Terblanche, who lived on a farm near Kromdraai, finally put Broom on the track of a very different sort of fossil man. Broom describes what happened as follows:

'On the forenoon of Wednesday, 8 June, 1938, when I met Barlow, he said, "I've something nice for you this morning," and he held out part of a fine palate with the first molar-tooth in position. I said, "Yes, it's quite nice. I'll give you a couple of pounds for it." He was delighted; so I wrote out a cheque, and put the specimen in my pocket. He did not seem quite willing to say where or how he had obtained it; and I did not press the matter. The specimen clearly belonged to a large ape-man, and was apparently different from the Sterkfontein being.

'I was again at Sterkfontein on Saturday, when I knew Barlow would be away. I showed the specimen to the native boys in the quarry; but none of them had ever seen it before. I felt sure it had not come from the quarry, as the matrix was different. On Tuesday forenoon I was again at Sterkfontein, when I insisted on Barlow telling me how he had got the specimen. I pointed out that two teeth had been freshly broken off, and that they might be lying where the specimen had been obtained. He apologized for having misled me; and told me it was a schoolboy, Gert Terblanche, who acted as a guide in the caves on Sundays, who had picked it up and given it to him. I found where Gert lived, about two miles away; but Barlow said he was sure to be away at school. Still, I set out for his home. There I met Gert's mother and sister. They told me that the place where the specimen was picked up was near the top of a hill about half a mile away; and the sister agreed to take me up to the place. She and her mother also told me that Gert had four beautiful teeth at school with him.

'The sister took us up the hill, and I picked up some fragments of the skull, and a couple of teeth; but she said she was sure Gert had some other nice pieces hidden away. Of course, I had to go to school to hunt up Gert.

'The road to the school was a very bad one, and we had to leave the car, and walk about a mile over rough ground. When we got there, it was about half-past twelve, and it was play time. I found the headmaster, and told him that I wanted to see Gert Terblanche in connection with some teeth he had picked up. Gert was soon found, and drew from the pocket of his trousers four of the most wonderful teeth ever seen in the world's history. These I promptly purchased from Gert, and transferred to my pocket. I had the palate with me, and I found that two of the teeth were the second pre-molar and second molar, and that they fitted on the palate. The two others were teeth of the other side. Gert told me about the piece he had hidden away. As the school did not break up till two o'clock, I suggested to the principal that I should give a lecture to the teachers and children about caves, how they were formed, and how bones got into them. He was delighted. So it was arranged; and I lectured to four teachers and about 120 children for over an hour, with blackboard illustrations, till it was nearly two o'clock. When I had finished, the principal broke up the school, and Gert came home with me. He took us up the hill, and brought out from his hiding place a beautiful lower jaw with two teeth in position. All the fragments that I could find at the spot I picked up.

'The spot where the skull had been found was much more carefully examined within a few days. All the ground in the neighbourhood was carefully worked over with a sieve, and every fragmentary bone or tooth collected.

'When all the fragments were cleaned and joined up, it was found that we had practically the whole of the left maxilla, the molar, and all the left temporal bone, with most of the left half of the sphenoid, and part of the occiput and parietal, with all the upper teeth except the canines and incisors and the third molar, but all these were represented by sockets. We also had nearly the complete horizontal ramus of the right mandible, with all the pre-molars and molars. As there was a satisfactory contact between the maxillary and temporal fragments, we could restore the side of the skull quite satisfactorily, except for the top, which had been weathered away.

'When the skull was restored, it was seen to be larger than that of the Sterkfontein ape-man, and to differ in a number of respects. The face is flatter, and the jaw more powerful. The teeth are larger, and in a number of characters different.'

'The Missing Link No Longer Missing'. That was how *The Illustrated London News* headed its account of the discovery on 20 August, 1938. Since in those days scientists felt that even relatively slight variations merited the invention of new types or even genera of early men, Broom called the Sterkfontein fossil *Plesianthropus transvaalensis*, and the Kromdraai man *Paranthropus robustus*.

Today we know that Broom's so-called 'Plesianthropus' is the same as the Taung child – it is an adult *Australopithecus africanus*. The Kromdraai man, on the other hand, is indeed another type, though not generically distinguishable from Australopithecus. These two forms, the smaller A-type and the more robust P-type, are basic to the discoveries of early humans in Africa. Their various and often entirely arbitrarily chosen scientific names caused some temporary confusion; but more about that later.

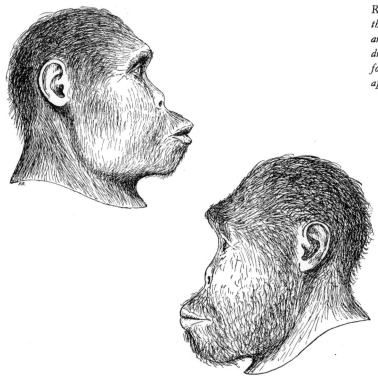

Robert Broom was already aware of the differences between the gracile and robust australopithecines. These drawings of his compare the Sterkfontein female (left), and Kromdraai ape-man (right)

The Second World War interrupted the excavation of the South African caves. Nevertheless Broom and his assistants were not idle. They sorted through vast quantities of animal fossils which had already been dug up, and thus managed to direct the experts' attention even more upon the African continent. Immediately war ended, General Smuts summoned Broom and, as the latter reports, declared 'that we were discovering the origin of man and that the works in the caves must be carried on'. The South African government offered him the necessary funds. Even though the majority of experts were as convinced as ever that the African finds were either not genuine hominids or that they were a species which had so to say by-passed man in their development, at least the work was now taken seriously, which had not been so at the time of the Taung discovery.

The australopithecine *Plesianthropus transvaalensis* ('Mrs Ples'): reconstruction model based on part of the female skeleton found at Sterkfontein

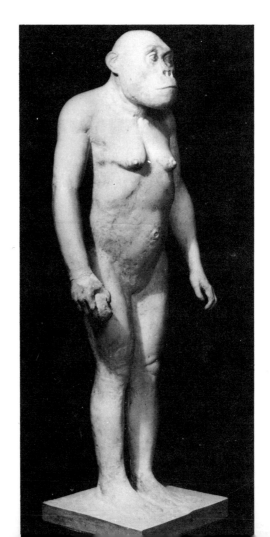

A decisive role in the change of general opinion was played by the two great prehistorians who were also churchmen, Henri Breuil and Pierre Teilhard de Chardin; both insisted on thoroughly inspecting the site of the discoveries and as a result were emphatic that Africa must be mankind's place of origin.

The age of the South African Australopithecus was in fact still much contested (the datings varied between five million and five hundred thousand years); nevertheless, very soon after war ended Teilhard de Chardin was prepared to risk stating his own opinion regarding the age of these finds. The South African australopithecines must, he decided, have lived anything from 500,000 to a million years ago, and in his view such evidence of their culture as had been found showed it to be identical to that of the Villefranchian period – that epoch which preceded the earliest European-Asiatic stone cultures. His comment on the finds themselves was: 'Their late Tertiary occurrence in Africa adds weight to the argument that this continent was the main birthplace of the human group.'

'Prometheus came from Africa' – this catch-phrase heralded Dart's fresh intervention in the search for Adam. Meanwhile, at Sterkfontein, Broom and his colleague John Talbot Robinson had brought to light the remains of twelve more australopithecines, among them the splendid skull of a well-grown female whom anthropologists soon nicknamed 'Mrs Ples'. She was the first australopithecine to cause something of a sensation in Europe and America. But Dart was excavating in a different area, in Central Transvaal,

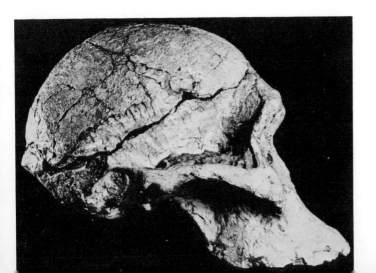

Side view of the skull of the now famous 'Mrs Ples'

in a stretch of land called Makapansgat (valley of Makapan) after a Bantu chief. Here he discovered not only numerous remains of A-type australopithecines but animal bones which in his view showed signs of fire, and what he thought were indications of ash in the ground. Dart believed these primitive creatures had already known the use of fire, and thus endowed Makapan man with the truly impressive name of *Australopithecus prometheus*.

It is now known that those remains of ash were probably manganese colouring. Nevertheless this error brought the australopithecines into the news, and enhanced their status even in the eyes of their original critics. Then Dart made a new discovery. Together with the remains of an australopithecine child aged about twelve whose jaw had been fragmented in a very peculiar way he found a whole lot of baboon skulls which had been similarly damaged. From this he concluded that not only Prometheus but also the biblical Cain had been active in Makapan.

'To kill one's own kind, roast and eat them is not the work of an ape, but of man', declared Hans Weinert, once so sceptical about the nature of the australopithecines. 'It was good to find the Prometheus act as the first achievement of developing man,' he added. 'But it seems we are obliged

Hypothetical reconstruction by Burian, of a community of australo-pithecines

to attribute the act of Cain to him as well.' In other words, in the view of Dart, Weinert and other anthropologists, the australopithecines had not been harmless eaters of plants and fruit, like modern anthropoid apes, but hunters, perhaps even cannibals. 'Killer apes', Dart called them.

A discovery which seems to fit in here was made by the young prehistorian C.K. Brain at Sterkfontein. He unearthed a number of roughly hand-size pebble tools made of quartz which had been chipped at one end in such a way as to form a sharp point or cutting edge. Robinson too discovered this sort of tool in South Africa. Some showed as many as fourteen facets from which flakes had been chipped off in order to perfect the shape. Inevitably critics wanted to attribute these ancient stone implements to later forms of man. It was only when the great discoveries were made in East Africa that something like a solution was found to the mystery of these pre-human stone artefacts.

Dart, however, was on a different track. Long before Broom's discoveries, during that period in his life when bureaucratic meddling had made Broom give up his excavation work in disgust, Dart's attention had been drawn to the strangely damaged baboon skulls found in the Australopithecus strata. They all looked as if the damage had been inflicted by a powerful blow. In 1946 Dart laid his collection of skulls before the eminent British anatomist, W.E. Le Gros Clark. Clark found them extremely interesting, and suggested Dart take the matter further.

For this purpose Dart enlisted the cooperation of an expert in forensic medicine with a detailed knowledge of the effects of violence, body damage and skull fractures, in the person of Professor R.H. Mackintosh, head of the department of forensic medicine at the University of the Witwatersrand. The verdict was that forty-two of Dart's baboons had undoubtedly been dispatched by a blow from a hard object. But these skulls came from the Australopithecus strata, representing a time when there were no real men living in Africa. Neither Dart nor Mackintosh believed that all the baboons could have been killed by a fall of rock.

Raymond Dart demonstrates how the shoulder blade of an ungulate can be used for killing prey

Bone implements of South African hominids

'A powerful downward, forward, and inward blow, delivered from the rear upon the right parietal bone by a double-headed object,' we read in one of the descriptions of the skulls, like a report issued from a morgue. Another runs: 'The V-shaped island of bone left standing above the obvious depression of the cranium shows that the implement used to smash it was double-headed . . . having vertical internal borders or sharp margins, and measuring approximately 30 mm between the two heads.'

Now Dart and Mackintosh needed to do some real detective work. They asked themselves the following questions: 1. What could this strange double-headed object have been? 2. Could the australopithecines have used such a thing, or have even made it? Dart was sure that the weapons which had delivered such blows must have been solid, long-lasting objects, definitely not perishable, and that they would be there if he looked for them in the *Australopithecus* strata.

And soon he had an ideal 'model' for such a weapon in his hand. In the breccia which had held the ape-men he found a surprisingly large number of thigh-bones of large and medium-sized antelope. A variety of precise tests were carried out which revealed that the knee-joint of this thigh-bone exactly fitted the mark of the blows on the baboon skulls. What is more, the same deposit contained a number of lower jaw-bones of small antelope still with a full complement of teeth. Such a jaw-bone, grasped in the hand, offered a human creature an ideal primitive saw or crude knife.

The same forensic tests were applied to the australopithecine child whose jaw-bone had been broken in such a strange way. Dart and Mackintosh experimented again and again with a variety of different skulls until they were absolutely sure that this particular damage could not have been caused by a falling stone or some similar accident. Obviously the young australopithecine had been killed for some unknown reason by one of his own kind.

When, in 1949, Dart published his conclusions in an article in the *American Journal of Physical Anthropology* entitled 'The Predatory Implemental Technique of Australopithecus', the response varied from icy rejection to sheer horror. The beautiful pseudo-Darwinian legend that man had gradually developed out of good-natured but dull-witted fruit-eaters into weapon-bearing masters of the world was so firmly rooted in the minds of scientists that few anthropologists could accept the idea that the early men of southern Africa could have been predators.

At a congress held at Livingstone in 1955 Dart met with the same reception as had Fuhlrott and other pioneers when they tried to convert conservative scientific opinion. Almost every prominent anthropologist journeyed to that African town near the Victoria Falls to participate in the gathering. Dart was allowed a mere twenty minutes to present his theory, which meant he could do no more than produce a cursory outline. The participants glanced at the extensive material he had brought with him, but dismissed it with a shrug of their shoulders, declaring that the bones had quite definitely not been collected by australopithecines but by hyenas who had carried them into the caves.

Australopithecus was a 'killer': he used bones such as this ungulate thigh-bone to kill wild animals and perhaps his own kind

Even Professor Leakey, so soon to be hailed as the discoverer of fossil man in Africa, believed neither in the use of weapons by australopithecines nor that this had any significance where man's family tree was concerned. He was of the opinion that during the Tertiary australopithecines had somehow cut themselves off from the species which was to develop into man, and had gone their own way. Dart's evidence he dismissed with the words, 'There does not seem to be any justification for regarding these caves as their dwelling places, and their bones, as well as those of the other fauna, were probably dragged into these caves by hyenas and other fauna.'

One of the most energetic opponents of Dart's views was von Koenigswald, discoverer and interpreter of Asiatic early man. Despite the African finds, von Koenigswald stuck to his opinion that Asia was the place of Adam's origin. Upon the subject of Dart's antelope bones, he wrote: 'It is easy to take such bones for implements, and this is in fact often done. But a comparison of the picture produced by Dart shows without any doubt that these bones have been gnawed and split by hyenas.'

The hyenas themselves ultimately disposed of the hyena theory. Some time after Broom's death in 1951, when Dart's cause appeared to be lost, the British palaeontologist A. J. Sutcliffe made an intensive study of hyena behaviour. The conclusion reached by Sutcliffe, after weighing up all the British evidence, was that hyenas could not possibly have committed the crimes with which they were charged, since the manner in which they use their powerful jaws to split or even crunch up the bones of their prey is a very distinctive one. Places where hyenas had fed would never contain whole thigh- or jaw-bones, but just a mass of pulverized bone – apart from teeth, foot bones and kneecaps. The South African breccia presented a very different picture: in the deposits where the remains of the australopithecines had been found there were bones which could have served these 'early men' as tools, just lying around, as it were, waiting to be put to use.

Thus Sutcliffe, with his first-hand knowledge of hyena behaviour, had no difficulty in disproving von Koenigswald's hypothesis; he had only to point out that when this animal goes to the trouble of dragging off its prey it does not leave bones lying around conveniently for archaeologists to find, but chews them all up, rejecting only the very hardest parts. Dart had once again been vindicated.

The German anthropologist Gerhard Heberer sums up this lengthy controversy as follows: 'The statistics produced by the study of many thousands of different sorts of bone fragments found near Makapan make it very evident that these fragments were collected together, used, and often even shaped for implements. Research has proved that hyenas could not have been responsible for such heaps of fragments. Dart first found much opposition to his interpretation, but the tenacity with which he put it forward and his assiduous furnishing of fresh evidence has finally met with success. Today scarcely any expert would seriously oppose him. The place near Makapan where the australopithecine discovery was made is now incontestably accepted

View of the mountain landscape of Northern Transvaal, from the mouth of the Makapan cave

as the lair of australopithecine hunters. A visit to the site is most rewarding, and an analysis of Dart's gigantic collections in Johannesburg entirely convincing.'

The regular manufacture of implements for a definite purpose is at the moment considered the most certain proof that the fossil creature in question was equipped with some intelligence. The australopithecines had reached this level. But it was only when prehistoric man had got as far as creating 'tools for making tools' that he had crossed the line which by modern definition divides the realms of animal and man in the real sense of the word.

The remains of more than a hundred australopithecines have been found in the stalagmite- and chalk-filled clefts and caves of South Africa since the days of Dart and Broom. Three years before his death Broom made a particularly remarkable discovery, though a much disputed one, near Swartkrans, about two miles from Sterkfontein. Here he found the remains of an australopithecine equipped with a particularly powerful jaw and truly nutcracker-like teeth. Later it was found that these 'Nutcracker men' even had a small sagittal crest on their skull resembling the crests of adult male gorillas and orangs. Broom gave these representatives of the P-type the provisional name of *Paranthropus crassidens* (large-toothed 'near-man').

The strong molars of the P-type were highly reminiscent of the Asiatic giant forms. Broom had already thought that the 'giants' of Old Java might be a related type. Robinson even considered the Javan *Meganthropus palaeojavanicus* to be directly related to the P-type Australopithecus. If this theory is correct then the australopithecines once inhabited parts of Asia too, perhaps evolving giant forms there. At the moment it seems as if this could indeed be the case; for the remains found of Javanese forms, scanty though they be, are older than the Asiatic early men but more recent than the African 'near-men'.

But despite the discovery of these many missing links, which fitted so neatly into their outline of man's development, Dart's and Broom's theory of a South African 'Garden of Eden' had not really become accepted even as late as 1959. Geographically South Africa seemed too far on the periphery of things: it was far more natural to think of it as an area to retreat to than as the cradle of man's progenitors. And this applies also to Java, should the Meganthropus type discovered there turn out to belong to the P-type australopithecines. Logically we would expect the area where man originated to be far more centrally situated – an area from which the progenitors of man could have spread gradually out into all surrounding regions.

At precisely the time these new doubts were being aired, Leakey's eighteen years of systematic excavation in East Africa were rewarded by the discovery of a site now rightly called the 'Gorge of the First Men'.

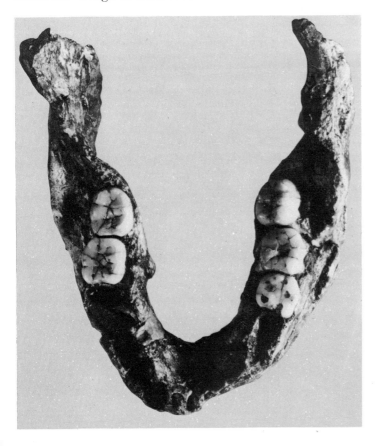

At Swartkrans were found not only the remains of robust australopithecines but also this mandible of Telanthropus capensis (see pp. 239– 240); some experts attribute the latter to a more advanced form of Australopithecus, while others consider it to belong to the Homo erectus group, that is, to early man

Olduvai is a word in the Masai language for the *san-sevieria*, a plant we like to grow indoors but which there grows wild. The name came to be applied to a deeply eroded gorge on the eastern side of the Serengeti plains in what is now Tanzania. That geological conditions in this area made it ideal territory for the archaeologist had long been recognized. Hans Reck reported in 1913: 'The Olduvai deposits are laid one upon another with great regularity, and consist of five main horizons, like five volumes in the enigmatic history of the earth laid out before our eyes. Each offers us entirely coherent documentary evidence concerning itself alone, and together they embody an unbroken sequence of the development of the earth, running like the thread of a story from the first to the final page. It is rare for strata to be so clearly distinguishable from one another as they are at Olduvai, the oldest at the bottom, the most recent at the top, undisturbed by a single gap, and never indurated or distorted by mountain-building forces'.

Reck was able to produce an impressive series of African animal fossils from the late Tertiary. But the apparently primitive man he excavated from Bed II turned out to be a Hamitic type who was not so ancient as he had thought. It

Leakey compares the reconstructed skull of a Zinjanthropus with the smaller skull of a chimpanzee

was a recent specimen obviously buried by his fellow men in a stratum of the earth to which he had not originally belonged.

Nevertheless this find attracted the attention of the young archaeologist Louis B. Leakey, who has since become the most notable scientist in East African prehistory. It was not, however, until 1931 that he started on the second Olduvai expedition. At Kanam and Kanjera he had found human remains associated with Acheulian tools. These according to some were very old representatives of *Homo sapiens*; others regarded them as close to Rhodesian man, while a third group maintained that they were African relatives of the Asian fossil men. But more important than these two rather confusing sites was to prove that gorge named after the plant *sansevieria*.

'The part of Serengeti into which the Olduvai gorge is carved was once occupied by a steadily sinking inland lake,' Heberer says of this classic site. 'It was gradually filled with material descending from the surrounding volcanoes, and this settled steadily, layer upon layer, in the area covered by the lake. Thus a splendid calendar of past history came into existence, the gorge today providing us with particulars of deposits hundreds of thousands of years old. It is a unique site for uncovering our earliest history.'

Leakey does not distinguish five deposits, or beds, as Reck had done, but four, which in their turn can be subdivided into a variety of quite clearly definable stratigraphical levels. Bed I, the oldest of these, is roughly 100 feet thick and consists mainly of lacustrine sediments. Here, as we shall learn, australopithecine remains were found which may well prove to be far older than those of their South African relatives. Bed II contains stone axes which have been evolved out of the pebble tools of Bed I. No prehuman or early human remains have been found in its lower deposits; its upper levels, however, produced remains of early *erectus*-like beings. In Beds III and IV tools were found which show very clearly how their weapon culture had developed. In Leakey's own words:

'Here in Olduvai we are confronted by a people who – in part at any rate – used raw materials which were just as difficult to work as those used by Peking man. Nevertheless their craftsmanship had developed to the same remarkable level of perfection as that of the men of the Stone Age cultures living in areas of England and France, who had flints of the finest quality with which to work.'

'On July 17, 1959, Dr. Leakey was unwell and remained in camp while his wife searched the slopes of Bed I, Olduvai,' writes Sonia Cole, describing the first major find in Olduvai's oldest layer. 'Tired and thirsty after several hours in the hot sun, she was about to return to camp when her keen eyes noticed a piece of bone in the process of being eroded out of the deposit. She recognized the texture of a skull, and closer inspection revealed the tops of some teeth. Excitedly she rushed back to fetch her husband, whose sickness vanished at once . . . It proved to be the skull of a young male hominid, broken up into tiny fragments that had to be pieced together like a jig-saw puzzle.'

Mary Leakey found innumerable fragments amongst the broken bones of predators and a number of primitive tools. These tools, made of water-worn pebbles, had obviously been fashioned by deftly striking one stone against another until it had acquired a relatively symmetrical cutting edge. These so-called 'pebble tools' are probably the implements used by the very first men. Furthermore, in the words of the primatologist John Napier, 'It seems possible that the crudeness of early tool-making may, at least in part, be attributable to the form and proportions of the hand itself.' Or, as Heberer puts it, the step from a tool-using chimpanzee to this primitive manufacturer of pebble tools can only be a small one, narrowing to nothing the gulf between animal and man.

After much trouble Mary Leakey managed to fit the 450 fragments of bone together to form a really very well-preserved skull. At first Leakey thought that this ancient Oldowan ancestor of ours was very different from the South African 'Nutcracker man', despite his powerful back teeth,

Opposite: map of the Olduvai Gorge, on the borders of the Serengeti Plain. Most of the discoveries were made where the main and side gorges fork. The diagram below represents a geological section – with elevation much magnified – through the Ngorongoro face and the Balbal depression. It also shows the geomorphological disposition of the successive faults which have occurred in the high levels of the Serengeti plain. F = fault

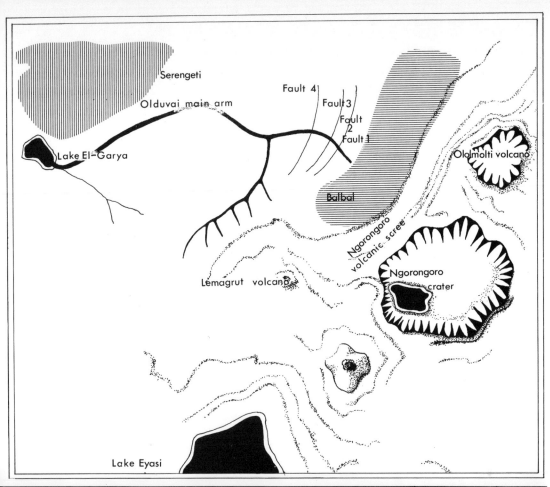

Serengeti

Olduvai main arm

Fault 4

Fault 3

Fault
2

Fault 1

Lake El-Garya

Ololmolti volcano

Balbal

Ngorongoro
volcanic scree

Ngorongoro
crater

Lemagrut volcano

Lake Eyasi

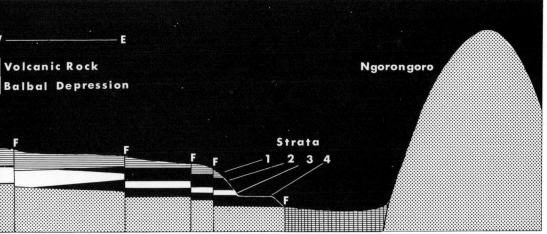

E

Volcanic Rock
Balbal Depression

Ngorongoro

F

F

F F

Strata

1 2 3 4

F

so reminiscent of the P-type. This was mainly because the shape of his face was so much closer to that of modern man. Therefore he was given yet another scientific name: *Zinjanthropus boisei*. 'Zinj' is an ancient name for East Africa; and the specific name is in honour of Mr Charles Boise, who was one of the small band of men who had first made it financially possible for Leakey to excavate in East Africa.

A large upper jaw fragment with several teeth, found on the shore of Lake Eyasi in the thirties, seemed to fit the Olduvai skull very well. This fragment was part of a collection which the German doctor and ethnographer Ludwig Kohl-Larsen had brought back with him from East Africa. Until the discovery of Zinjanthropus it was thought to be an African counterpart to the Asiatic 'giants', and was known as *Meganthropus africanus*. But now most experts look upon him as a robust australopithecine.

But before Leakey could delve more deeply into the secrets of the Olduvai Gorge, two finds were made in other parts of Africa. In 1961 the French palaeontologist Yves

Frontal view of a reconstructed Zinjanthropus skull, showing the sagittal crest

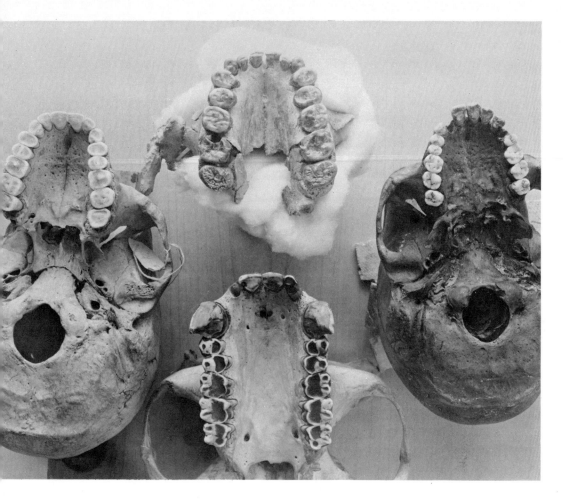

Coppens and his wife discovered a skull fragment with parts of the face at Yayo near Koro Toro, roughly a hundred miles north-east of the Lake Chad. Three years later a well-preserved lower jaw was dug up near Peninj on the shore of Lake Natron in East Africa. These two fragments, both in dentition and proportions, seemed to fit Leakey's Zinjanthropus splendidly. Later the Koro Toro find was reckoned not to have any connection after all; it seems more likely to have belonged to an early man of the *erectus* type. Yet Zinjanthropus remained, and with it the question of its age.

Palate of Zinjanthropus boisei (top), compared with those of an Australian aborigine (left), a modern European (right), and a gorilla (bottom)

In July 1959, Leakey extracted the two halves of the Zinjanthropus upper jaw from the floor of the Olduvai Gorge, with an instrument such as dentists use

Attempts to date African fossil man now went back into the Tertiary. The South African scientist, C.K. Brain, had devised a very ingenious method. Geologists had known for a long time that during the final stages of the Tertiary the climate of Africa had been extremely dry, but that at the start of the Pleistocene it had turned wet. Brain therefore studied the sand of each deposit in which australopithecine remains had been found. If the ratio of quartz in the sand grains was high compared with that of chert crystals, then the climate must have been dry at the time; if the grains of sand were rounded, this must have been caused by high winds – also a sign of a dry climate. Brain found all this applied to the composition of the deposits he studied. Were they, therefore, still Tertiary?

A more exact dating was achieved by the scientific use of the 'Earth clocks' – by the potassium-argon method previously referred to. Argon is a very rare element; it is to be seen, together with the reddish neon, as the blue gas emitted by illuminated signs. Since a radioactive isotope of the common element potassium, K 40, decays after an enormously long half-life into Argon 40, radioactive decay must have taken place if argon is found in a crystal of potassium. Hence the more argon there is, the older the deposit. In theory this means that the potassium-argon test can be used to ascertain the age of finds dating back well beyond the range of Carbon 14.

Louis and Mary Leakey at work in the Olduvai camp, putting together the two halves of the Zinjanthropus upper jaw

While the American scientists J.F. Evernden and G.H. Curtis were applying this test to Zinjanthropus, Brain's findings duly caused yet another sensation. On 23 July, 1961 he announced that he could prove the Oldowan man to be not less than 750,000 years old. Thus there was now no doubt that Leakey had found the earliest hominid yet, appreciably pre-dating the South African A (gracile) and P (robust) types. However, he was certainly not a true 'Tertiary' man, for it is generally held today that the Tertiary ended as much as two or even three million years ago.

Kenneth P. Oakley, who proved that the Piltdown skull was a fake, visiting the site of the Sterkfontein discovery in 1953

The man who had exposed the Piltdown forgery, Kenneth Oakley, had hitherto questioned not only Dart's theory of 'hunting killer-apes' but also whether Australopithecus could really be called human. This clear-headed and to all appearances very reserved Englishman has for years been deemed the good or sometimes even the bad conscience of anthropologists, often preventing his colleagues from arriving at over-hasty conclusions.

In 1956, that is, even before the epoch-making Olduvai discoveries, Oakley accepted the evidence which had come out of South Africa: 'It is agreed by the majority of anthropologists that the australopithecines were either part of, or very close to, the line of evolution that led to man ... It is of course highly likely that pre-human hominids were semicarnivorous and that they made use of stones, sticks and bones as ready-to-hand weapons and tools; but to prove it is difficult.'

Thus a conclusive verdict had been reached by what might be called anthropology's highest seat of judgment. With the exception of a few who still stuck to their belief in Asia as the earliest home of man, almost the entire scientific world now believed Dart's and Broom's 'African creation story'. The fact that Leakey had plumbed far greater depths of past history in his Olduvai excavations merely hardened the evidence – to use the language of the courts. Nevertheless it would have been very difficult to prove that a genuine member of the genus Homo, an intelligent being with

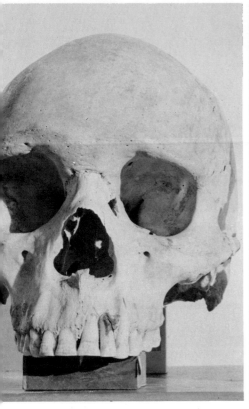
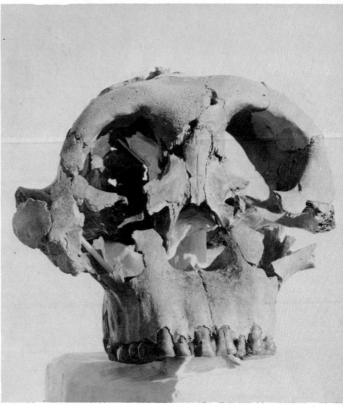

mastery of his surroundings, could have arisen in so comparatively short a time out of African australopithecines whose brain capacity was scarcely as large as that of an anthropoid ape. But in South Africa the first hints of a solution to this apparent problem had already been found.

An oddly human-looking piece of palate and lower jaw resembling that of Heidelberg man, as well as a further very human-looking bone, had already been dug up during 1952 and 1953 by Robinson in the neighbourhood of Swartkrans. Did this *Telanthropus capensis*, as his discoverer called him, belong to the australopithecine group? Did he represent a further evolutionary development of the *erectus* type towards early man? Perhaps, as a descendant of the australopithecines,

Skull of Zinjanthropus (right), compared with the particularly large-toothed skull of an Australian aborigine. The aboriginal inhabitants of Australia are, anthropologically, among the most primitive races of men still living today

239

he even represented that root from which had stemmed not only early man but also the direct forerunners of *Homo sapiens*?

Leakey and his colleagues approached this problem by looking at the primitive stone tools, those pebble tools brought to light in such numbers from the Australopithecus deposits in East as well as South Africa. Sonia Cole writes: 'The question now arises: Which of the hominids, *Australo-*

The australopithecines will have used pebble tools such as these for skinning the animals they caught

pithecus africanus, Australopithecus robustus, or *"Telanthropus"* was the toolmaker? Or were all three toolmakers? It seems unlikely that Australopithecus made tools in his early stages, during the drier climatic phase, or some would surely have been discovered in the [Makapan] Limeworks Cave midden. By the time of the succeeding wetter phase ... he may have learned this art; or of course there is the possibility that the pebble tools there were made by another kind of hominid, although their remains have not been found at this site. It seems very likely that the pithecanthropine "Telanthropus" was the maker of the Chellean-type tools at Swartkrans, for we would expect Australopithecus, if he made tools at all, to make Oldowan pebble tools only.'

Thus the first australopithecines, whether of the gracile A or robust P type, were – to use Oakley's definition – merely tool-users, but Telanthropus (who is at present known by the scientific name *Homo erectus capensis*) had already become a sophisticated tool-maker. This is also suggested by a find which Leakey made in 1960 in the slopes of Bed II at Olduvai, from which he washed a really large skull, 209 mm. high and 150 mm. broad, with a vault less high than that of Steinheim man but loftier than that of Peking man. The brow-ridges were particularly strongly developed; such thickness and such prominence had not yet been ascribed to a member of the genus Homo. *Homo erectus leakeyi*, as he was called after his discoverer, was shown by the potassium-argon test to have an absolute age of at least half a million years. Thus modern dating methods made this 'Chellean man' of Olduvai one of the oldest early men to have been found.

This Oldowan early man was taken to be related to *Homo erectus*, a supposition supported by the typical Chellean tools found in the same stratum. These are far better made than the pebble tools, which had hitherto been associated with the australopithecines. Nevertheless it seems likely, in the light of the finds from Swartkrans and Olduvai, that the direct descendants of Australopithecus went through an *erectus* stage, till they branched off into the Asiatic type of

Hypothetical representation of Australopithecus

early man on the one hand and into the direct forerunners of *Homo sapiens* on the other. For, as the American palaeontologist George Gaylord Simpson has stated, there is a strong suggestion 'that much of the rectilinearity of evolution is the product rather of the tendency of the minds of scientists to move in straight lines than of a tendency for nature to do so.'

The two types of Australopithecus, with and without the sagittal crest. Left half, gracile type, with more powerful front teeth but a less heavy facial skeleton than the robust type (right half)

The transition from pre-human to genuine human beings was now no longer an unbridgeable chasm. Yet the question whether Australopithecus had not only used bones as tools but also made stone tools himself remained unanswered. In fact only the A type (that is, creatures known under the scientific names of *Australopithecus africanus, Australopithecus prometheus* and *Plesianthropus transvaalensis*) would appear to have been tool-makers, the makers of the much-discussed pebble tools. The more robust P type (under which category anthropologists include the South African *Paranthropus robustus* and *Paranthropus crassidens*, the east African *Zinjanthropus boisei*, and also the very diverse *Meganthropus* types) cannot, it seems, be classed as such.

Some very fundamental differences between the two types bear this out. The front teeth of the P type are small, scarcely fitted for the job of meat-eating. The large nut-cracker-like back teeth, moreover, necessitate a large jaw, a strong muscular system for chewing, and a sagittal crest for attachment of the temporal muscles. All this points to the probability that the large, heavier P type was a specialized plant-eater, requiring no tools. Today its representatives are known by the scientific names *Australopithecus (Paranthropus) robustus* and *Australopithecus (Zinjanthropus) boisei*.

The smaller A type, on the other hand, who more closely resembles modern man and who is now classed as *Australopithecus africanus*, possesses a set of front teeth admirably suited to tearing flesh. Also, his molars are smaller, and he lacks the strong jaw and sagittal crest. His dentition suggests that, like modern man, he could eat virtually anything.

It is easy to imagine him already able to hunt and capture prey. But one thing common to both is that their canines do not project above the level of the other teeth, as they do in the anthropoid apes. Also, they lack the monkey gap, or simian diastema, and the so-called 'simian shelf'. Heberer sums it up as follows:

'It looks as if the two Australopithecus types might have lived in different surroundings and therefore had different sources of nourishment. In this case they could perhaps have existed simultaneously alongside each other. Whether the two groups were already genetically isolated or whether they still interbred and so exchanged genes is not known. The A-type went on to evolve into more advanced types of men, whereas the P type probably became an offshoot with no further historical future.'

Leakey and Heberer also point out that teeth are changed at a much later stage in the life of the australopithecines than in modern anthropoid apes: 'One can reckon that, unlike anthropoid apes, they will have had a prolonged youth, as do humans. This means the australopithecines had a longer "learning period" than anthropoid apes. Such retention of the attributes of youth is called "Neoteny" in biology. It is a decisive part of the process of becoming human.'

The next announcement from Olduvai came in February 1961. Leakey's son Jonathan found some fragments of bone in a still more ancient deposit of Bed I. These were temporarily christened pre-Zinjanthropus. They appeared to belong to a juvenile australopithecine, but oddly enough he seemed to have possessed a slightly larger brain than those apparently descended from him. But at first the scientific world was only interested in the enormous age of this new find, and also in Leakey's assertion at a Washington press

Hand bones of the Homo habilis from Olduvai, believed to be 1·75 million years old

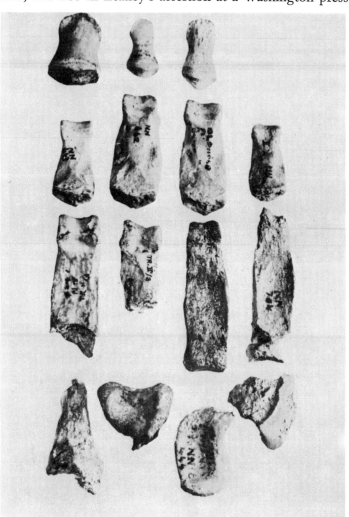

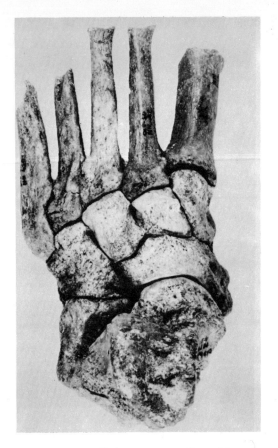

conference that this might well be a case of prehistoric murder. The bone fractures, like those in some South African australopithecines, seemed to point to violent death.

Between then and 1964 the remains of four other individuals came to light in the same deposit: a skull and parts of the upper and lower jaw of a woman roughly twenty years old who must have been between 3 feet 6 inches and 4 feet 6 inches tall, as well as the remains of hands and feet which looked perplexingly human. The hand belonging to a pygmy seemed advanced enough to have easily fashioned tools and to have broken those animal bones found with it in the deposit. Leakey carefully studied this amazing new find, estimated to be nearly 2 million years old, and in April 1964, together with his colleagues, published a fully illustrated article in the

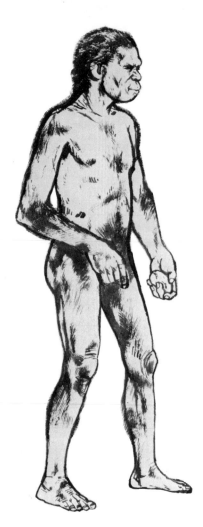

Barry Driscoll's drawing of the habilis type. Perhaps he is a link between ape-man and genuine man

distinguished journal *Nature*, apparently completely reversing all theories so far put forward, and culminating in the statement that the Oldowan pygmy is a member of the genus Homo, that is, a true man, and yet 2 million years old; hence even older than the australopithecines which till then had been deemed ancestral to man.

This discovery placed the beginning of the human race three-quarters of a million years further back in the past, wrote Leakey daringly. The text books had to be rewritten – including one of his which was in the process of being printed.

Homo habilis was the name Leakey gave the new find. By calling him 'skilled', or 'dexterous', he meant to stress his manual skill. It is not difficult to imagine how hotly opposed most anthropologists were to this new theory. Here they were presented with yet another theory making out the genus Homo to be older than his presumed ancestors. And that it should be one of the most distinguished of living authorities on the subject of east African prehistory who was thus prepared to shatter all prevailing views, made it all the worse.

Even so, Leakey had staunch supporters in his distinguished colleagues Philip V. Tobias and John Napier. Tobias, a respected anatomist and brain specialist, estimated the brain capacity of this juvenile skull to have been around 680 c.c. This made it almost 80 c.c. more than the large-brained Australopithecus but still almost 100 c.c. less than the smallest-brained members of our own genus (i.e. *erectus*). Hence it was apparently a transitional form, that is, one which could not be interpreted as belonging to either category. And there in limbo it remained, till anthropologists were prepared to widen their criterion of brain capacity and be a little less rigid in their definition of the dividing line between the genera Australopithecus and Homo.

'One cannot measure intelligence in cubic centimetres,' the old maestro Charles Darwin acutely observed. Heberer added that it was risky to place the habilines in the genus Homo: 'Perhaps it would be better not to extend the range of capacity rate down to 680 c.c. and below, but instead to raise it up to this level for the australopithecines. Then the habilines would belong to the genus Australopithecus.' A comparison of the skull contours substantiates this view: 'If one places the mid-line of the Australopithecus skull from Makapan against that of the habilis skull from Olduvai it requires no special anthropological knowledge to see that the outlines are almost identical.'

The question whether the habilines should be considered a group in themselves is still an open one for whereas those found in Bed I of Olduvai seem to be more closely related to the A type australopithecines, others from the lowest deposit of Bed II have a closer resemblance to the *Homo erectus* group. But such details concern the experts only. The discovery of the skilled pygmy of Olduvai has not, as Leakey thought, turned current opinion upside down, but on the contrary has provided us with yet another link in the chain of man's development. Here again we can see that evolution did not take place through a series of adventurous 'leaps', but by means of a gradual, sometimes almost imperceptible progression. As Heberer says: 'The gulf between animal and man is not great – neither is that between genera Australopithecus and Homo.'

The fact that smaller, more intelligent and less specialized hominids already existed alongside or even before their more robust generic comrades surprises no one who remembers that the evolution of species does not follow the lines of a straight-branched genealogical tree, but rather those of a multi-dimensional and widely ramified bush. The line of development running from the habilis type, who had already invented tools and hunted his animal-like contemporaries, to the stone-working early men living together in hordes or even in well organized communities, and from them to the men of the late Pleistocene with cult practices and art is now shown to be practically unbroken.

It says much for the sagacity of old Robert Broom that he wrote so tellingly about pre-human development long before the Oldowan discoveries: 'To have lived in caves and

Man

Chimpanzee

Pygmy Chimpanzee

Gorilla

Orang-Utan

hunted and killed baboons, to have dug out moles and spring-hares, to have been able to capture young antelopes, all show a very considerable degree of intelligence, greater than is found in any living being except man. They were so nearly men that I think we are safe in calling them ape-men.'

Logically, however, australopithecines must be descended from anthropoid apes. For want of any human-like ape fossils the anthropologists of Darwin's time had been forced to study the extant anthropoid apes – for instance the chimpanzee, which would appear to be most closely related to man; the gorilla and orang-utan, at one time very much the favourites; and finally the six types of gibbon which Ernst Haeckel had reckoned to be quite near the common root of the families of anthropoid apes and man. Yet the Western ape-complex is so deeply embedded in our society that up till some two decades ago it was regarded as highly suspect to accept them as our closest relatives. The higher primates, despite their popularity in zoos and menageries, were still regarded as rather strange forms of life, hardly to be taken seriously; mere caricatures of man. The moralists for their part mistrusted them, finding their often all too human characteristics a source of embarrassment.

Even after serious primatological research had begun, prudery and the ape-complex were still so prevalent in civilized society, that one of the first men to study the behaviour of apes, Sir Solly Zuckerman (and he only studied

Opposite: A comparison of the chromosome series of the higher Primates. The similarity between the human chromosomes and those of chimpanzees and pygmy chimpanzees is conspicuous, and suggests a common ancestry.
Below: A comparison of the facial muscles of man and the chimpanzee, revealing a very considerable difference

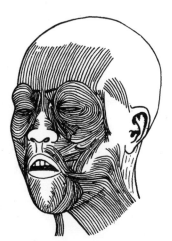

them in zoos and so arrived at some very false conclusions) had to go through some trying experiences as a result of the general prejudice he found existed against these animals. The story goes that when he wanted to publish his book under the title *The Sexual Life of Primates*, his London publisher delicately explained to him that in England the word 'primate' is generally understood to refer to a dignitary of the Church of England. Zuckerman, South African born, was unfamiliar with the refinements of the English language. In the end he was persuaded to publish his book under the title *The Social Life of Monkeys and Apes*.

In such a climate of prejudice, misunderstanding and false shame an uninhibited and objective comparison of the behaviour of anthropoid apes and man was a tricky subject to tackle. Today still we have not yet wholly rid ourselves of these complexes. Thus, when in 1967 the zoologist Desmond Morris published a book in which he attempted to show that human behaviour is derived from our ape ancestors, it created something of a sensation.

Morris had previously written a book on the art of chimpanzees, and this too had proved controversial; the fact that anthropoid apes can draw in charcoal and paint in colour is

The male orang-utan 'Alexander' at the easel. A painting done by the young chimpanzee 'Congo' is on the right: it shows the typical fan shape of a chimpanzee's brush strokes

another thing which all too many self-assured persons seem to find an affront to their ego. For on the face of it chimpanzee paintings bear a devastating resemblance to those works of art that adorn international art exhibitions. Eminent critics compare the pictures painted by monkeys with tachism, or action painting; indeed, Picasso was so impressed by one of these artistic efforts of a chimpanzee, that he acquired it for himself. The most logical and reasonable conclusion so far drawn from the fact that monkeys are able to wield paint brushes is that of the German zoologist Bernard Rensch: 'The animals regard this activity as a game in which their delight in the mechanics of it is often the driving force.'

Rightly or wrongly, opponents of modern art insist that the works of recent abstract painters are directly comparable with pictures painted by monkeys. Be that as it may, Morris does clearly establish something we can already detect in Ice Age art, namely that what we call art is definitely an inherited talent acquired very early on in our development – a biological necessity, as it were.

In 'The Naked Ape' Morris aims to demolish the prevailing taboos we have erected to protect us from our embarrassing kin. Morris maintains that man had already lost his coat of hair in prehistoric times as a result of his active way of life as a hunter. This brought about fundamental changes in his make-up; for if he was to survive, he had to become aggressive, to develop into a predator. The upright posture caused a very early re-orientation of female sexual signals, these being transferred from the rear to the front of the body. Sexuality came to be expressed through breasts and lips, and the tendency towards pair bonding which established itself increasingly resulted in the custom of kissing and other personal manifestations of affection.

The conception of our early history with which Morris starts his thesis is a very compelling one; it is shared in great part by those who have been exploring the problem of African prehistory. In the Miocene, that is roughly 25 to 20 million years ago, the African forest gradually dwindled, reaching its lowest extent by the time of the Pliocene, the final epoch of the Tertiary. The savannah vegetation spread over a large part of the African continent. Later, in the Pleistocene, the period coinciding with our European Ice Age, the forest gradually reasserted itself over some parts of the continent. In the course of the Pliocene period, which lasted about 15 million years, many African primates found themselves compelled to give up forest life and take to the open savannah; many sorts of anthropoid apes began to walk upright during this long era, which, anthropologists believe, also saw the emergence of man.

This all sounds very plausible, and accords completely with the African Pliocene finds of hominids we shall soon describe. For Morris thinks that the ancestors of the modern anthropoid apes remained in the jungle, since when their numbers have slowly and steadily decreased. But those apes which were able to adapt themselves to living on the open plains withstood the competition of the other inhabitants of the savannah. They could organize their meat supply better than the native beasts of prey. They developed a large, able brain. With their superior sight and their manual skill they became more and more removed from their cousins in the jungle, evolving along very different lines. They began to fashion tools and weapons; their societies became firm communities. They possessed and defended particular territories of their own, discovered the use of fire, and made themselves secure dwelling places. They became 'predatory' apes; but also the first to initiate cultural activities and hand them down.

A tool-using ape: this five-year-old female chimpanzee, having selected a grass stalk, uses it to dig out termites, a favourite food

'Angel of the chimpanzees' they called Jane van Lawick-Goodall, the young British zoologist, who worked for some time as Leakey's secretary in the Olduvai gorge. Unlike Morris, she possesses first-hand knowledge of chimpanzees living in the wild. Many other primatologists too have during the past decade been collecting a vast amount of material in the field about anthropoid apes. The data they assembled made it very clear that Morris seriously underestimated our 'backward' forest cousins.

Those such as Jane van Lawick-Goodall, Frankie and Vernon Reynolds, Adriaan Kortlandt and others who studied animals in the wild found that chimpanzees use small sticks or straws of grass as a means of capturing termites. Also, they cut themselves sticks to enhance the usefulness of the natural objects they use as tools. Nor were they predominantly unsophisticated plant-eaters, as Morris had thought: savannah chimpanzees living on the open plains destroy and consume mammals as large as a bush pig or small antelope. They too had most certainly taken to the savannah, which Morris thought had been the life adopted by the naked ape alone. The fact that they are now usually found in the jungle is mainly because on the plains they are no longer able to compete with man, an aggressive creature with a vastly superior knowledge of weapons.

Jane van Lawick-Goodall makes friends with an eleven-month-old wild chimpanzee baby 'Flint'

253

The degree of intelligence shown by chimpanzees living in captivity has for several decades past been a subject of scientific study, and the results have been even more surprising. In American experimental stations chimpanzees have been taught to use automatic machines, and have even managed to learn the 'buying power' of different coloured coins. One coin, they find, causes the machine to issue them with small helpings of food, whereas another produces more. Their hunger satisfied, so the American zoologist John B. Wolfe tells us, some even 'hoard' their coins. Whereas other intelligent animals 'think while they act', as Konrad Lorenz puts it, experiments have shown that chimpanzees and other great apes behave in such a way as to suggest that they rely very little on inherited ways of behaviour, but follow a genuine 'tradition'.

Lorenz has described the actions of a great ape intent on getting hold of a banana dangling from the ceiling. First he looked at the banana, then at a box which happened to be in the same room, scratched his head like a man deep in thought and was obviously considering the problem. 'The matter gave him no peace, and he returned to it again. Then suddenly – and there is no other way to describe it – his previously gloomy face "lit up". His eyes now moved from the banana to the empty space beneath it on the ground, from this to the box, then back to the space, and from there up to the banana. The next moment he gave a cry of joy, and somersaulted over to the box in sheer high spirits. Completely assured of his success, he pushed the box below the banana. No man watching him could doubt the existence of a genuine "Aha" experience in anthropoid apes.'

The Berlin animal psychologist Gunther Tembrock adds: 'The chimpanzee also plays, he also discovers tools and their function; thus in him too there must be a dawning awareness of self.'

'Julia', a female chimpanzee, deftly fits the right key into a small padlock, a device she had never seen before

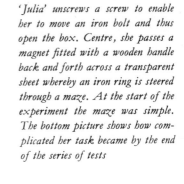

'Julia' unscrews a screw to enable her to move an iron bolt and thus open the box. Centre, she passes a magnet fitted with a wooden handle back and forth across a transparent sheet whereby an iron ring is steered through a maze. At the start of the experiment the maze was simple. The bottom picture shows how complicated her task became by the end of the series of tests

A chimpanzee enters a plantation, after the much coveted pawpaw fruit. He stands erect, thus giving himself a better view and advance warning of any potential dangers

'Dehumanized apes' was the description the Dutch zoologist Kortlandt applied to our chimpanzee relations after studying them in their natural habitat. The question he asked himself was this: where did chimpanzees acquire the perplexing capabilities they exhibit in captivity but seldom in the wild state? Answering Morris, Kortlandt put forward the hypothesis that all chimpanzees originally lived in the savannah. His theory, he maintains is supported not only by the hunting habits of these animals but also by their typical movements when attacking: they usually attack in the upright position, and have even attempted to kill manipulated dummy leopards by means of sticks and clubs – sometimes with a sureness of aim previously attributed at the most to the australopithecines.

Long bipedal runs are not practical in the jungle; but in the savannah they make good sense, for in the upright position a creature can see further above the level of the grass. And taking up a cudgel to throw at your enemy often achieves a better effect than does the straightforward attack of a beast of prey. The cultural and physiologically behavioural difference between chimpanzees and australopithecines is in Kortland's opinion so small that one cannot really talk of a gulf between animal and man. Also, the 'reorientation of sexual signals' which Morris asserts took place has developed in the great apes to such an extent that pygmy chimpanzees, occasionally orang-utans, and in rare cases even common chimpanzees will mate frontally.

According to Kortlandt's dehumanization hypothesis chimpanzees and perhaps also gorillas are in fact 'creatures of culture' – 'culture' being interpreted in its widest sense as behaviour learnt in youth and later passed on to their progeny. Faced with the superior australopithecines, with their bone cudgels and pebble tools, the apes fled into the jungle, despite their nearly comparable abilities and intellectual equipment.

Thus the old theory that the progenitors of pongids and hominids were derived from the same root has been confirmed by modern behavioural research. These common ancestors will now be considered.

Left: This female chimpanzee was caught in the savannah and now lives in a state of semi-wildness in a camp. She is here seen attacking a leopard dummy with a six-foot-long stick, which she brings down on her enemy with lightning speed

Below: A forest chimpanzee, seeing a leopard dummy, breaks off a bough, tears off the leaves, thus providing himself with a cudgel. He does not, however, attack the leopard immediately, but hesitates, lowering the stick. Were he a savannah chimp, on the other hand, he would not hesitate to club a 'leopard' of this sort, and would make a thorough job of it, even 'beheading' it with the force of his blows

A fossil genus of 'forest ape' of the Miocene and lower Pliocene was for a long time considered the 'model' for the common ancestors of anthropoid apes and those apes on the way to becoming man. Remains of Dryopithecus, as this creature was named, had been found as early as the nineteenth century. A thigh bone excavated from the Eppelsheim sands near Mainz looked so human that the philosopher Schleiermacher thought it must belong to a human child. Then in 1856 Edouard Lartet found a bone from the upper-arm and some fragments of skull near Saint-Gaudens, on the northern slope of the Pyrenees. Also the Württemberg deposits produced many teeth which for some time were believed to belong to some prehistoric deer, or a pig or even a man. Discoveries made in the twentieth century showed, however, that Dryopithecus had also lived in Asia and Africa. Whether he really was a 'forest ape', as his scientific name implies, or whether he inhabited the plains, remained an open question until Leakey's research.

In the words of Carus Sterne, Dryopithecus was a creature 'closer to the ancestral line of man than of any living ape'. The French palaeontologist Albert Gaudry even went so far as to classify this creature, not as a great ape but as an 'animal-like forest-man', able to shape tools from stones by chipping. However, he had later to abandon this hypothesis. Even in 1945 Roy Chapman Andrews declared that Dryopithecus had possessed the potential to become a man which modern great apes never had, and was therefore in a position to bridge the gap between animal and man.

An island in Lake Victoria soon afterwards provided us with fresh evidence. Leakey had come across this site by pure chance in 1926. The night steamer by which he was crossing the great East African lake happened to be late, so he saw Rusinga Island through binoculars for the first time in full light of day. His practised eye could see it was a promising place to look for fossils.

But it was only in 1931 that he was finally able to take his team of assistants there. On the very first day of excavation he found the fossil jaw of a Miocene ape. A. T. Hopwood, then a Curator at the British Museum, thought it was a fossil chimpanzee and gave it the name *Proconsul*, after Consul, a popular chimpanzee in the London zoo. But gradually it became apparent that about 20 million years ago the island of Rusinga, a mere nine miles by five, must have been a real paradise for apes. Leakey and his scientific assistants discovered no less than three species and six types of ape there, ranging in size from that of a gibbon up to that of a gorilla, all in all more than a hundred specimens. The most important of these seem to be Proconsul, three species of this type alone being found. A particularly splendid skull of *Proconsul africanus* type was excavated by Mary Leakey in 1948.

These studies revealed not only that Proconsul was older than the European Dryopithecus but that he also looked far more like a hominid. His arms were shorter than his legs, which is not the case with modern apes. Thus this

Leakey at work on Rusinga Island

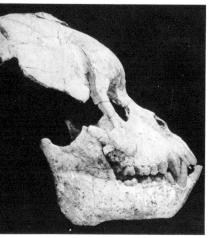

Above: the skull of Proconsul, frontal and lateral view
Below: Diagrammatic reconstruction in section of Proconsul's head and neck

creature had not spent all his life swinging from tree to tree but part of it on the ground. His orbits were particularly remarkable, for they differed from the round apertures typical of the ape skull, and were more like the rectangular sockets of man. Also, the thumb of Proconsul was developed as in man, not shortened as in pongids. But in other ways this Middle Tertiary Rusinga ape seemed more at home in the realm of the pongids. All in all he seemed an ideal ancestral model for the common forefathers of pongids and hominids.

Many scientists still believe Proconsul to have played this role, but others believe his great dagger-like teeth rule him out of the ancestral line of man. In any case he was an early ape of the Lower Miocene who had no supraorbital ridges, no sagittal crest and no pronounced snout. The Proconsul apes probably lived an agile life on the savannah, loping along on all fours through the open country from one group of trees to another, but perhaps able to advance half upright when the need arose. 'But the upright posture is a precondition for the development of the brain into a human one,' writes Heberer of Proconsul. 'Once this had happened, the first decisive step in the process of transformation had been taken, at the end of which the creature would cross the rubicon between animal and man.'

The Mountain Ape of Monte Bamboli seemed at one time to invalidate the now generally accepted family tree of man, by which the gibbon-like *Propliopithecus haeckeli* of the lower Oligocene is thought to have evolved into Proconsul-like forms and through these into the modern anthropoid apes on the one hand and australopithecines and early men

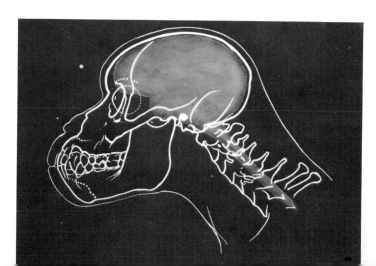

on the other. Paul Gervais, the French palaeontologist, had found the remains of a lower jaw in a lignite pit on Monte Bamboli in the Italian province of Grosseto back in 1872. He attributed it to a long-tailed monkey-like creature of the late Tertiary. It was given the name *Oreopithecus bamboli*, 'Mountain Ape of Mount Bamboli', and its age put at about 10 million years. For three-quarters of a century this mountain ape attracted little attention. True, the English scientist Forsyth Major pointed out the remarkable similarity of its jaw to that of a human, but no one took much notice at the time.

Then in 1949 the palaeontologist Johannes Hürzeler of Basle made an intensive study of the Oreopithecus remains to be found in the collections in Paris and Florence. Five years later he published a sensational report about them: the dentition of these 'apes' could not in fact have belonged to an ape at all but to a creature on the way to becoming man – or, more precisely, a great ape with far more human characteristics than Proconsul ever possessed. Large-scale excavations were then carried out on Monte Bamboli, financed by the Wenner-Gren Foundation and Swiss sponsors. By 1958 further Oreopithecus remains had been uncovered, and then finally a complete skeleton, discovered 810 m. down. As the pressure of earth from the mountain was very heavy they had great trouble in extracting the skeleton; but it turned out that the real difficulties had only just begun. The lignite pit of Baccinello in which Hürzeler found the bones had become uneconomic and was forced to close, and Hürzeler and his colleagues were unable to raise the funds to buy it. Moreover, the Italian authorities refused to allow the skeleton to be taken to Basle. This was because a report had spread through the world press that Hürzeler had found 'a man ten million years old', and was thus upsetting the prevailing theories of the origin of man.

A man before mankind evolved – this was how the mountain ape of Monte Bamboli was designated for many years. Other scientists took it to be a transitional form between a true ape and a great ape. But more recent investigations have shown that it cannot have been a hominid, despite the fact that it had some features in common with man, some with apes and yet others with cercopithecoids. Today, together with *Apidium*, the fossil primate of Oligocene Egypt, the Monte Bamboli ape is classed as a distinct family.

'Certain indications of specialization mark it out as an offshoot,' writes Heberer. His verdict hit the headlines for a while. 'It was definitely not one of our forebears, but very likely a sub-human ape, an example of an early form of hominoid. Not only does it prove that the hominoids were already independent as a group as much as 10 million years ago, but also that they are not descended from long-armed swinging climbers of the tropical rain forests; for at that time, that is, 10 million years ago, it is highly unlikely that long-armed apes had already come into existence.'

But since 1962 these 10-million-year-old creatures can no longer claim to be our oldest ancestors; this role has been

taken over by hominids who in their original form lived around 14 million years ago and who seem to bridge the gap between Proconsul and Australopithecus, that is, the gap between hominoids and hominids. Today, provisionally, they mark the most distant point to which we can trace our ancestry.

Oreopithecus was not a direct ancestor of the human race, but a separate branch on our family tree

At a Washington press conference in 1962 Leakey announced the discovery of two fragments of the lower jaw of a hominid found at Fort Ternan in west Kenya. He named his discovery *Kenyapithecus wickeri*, after the botanist Fred Wicker on whose land he had made the find. Sonia Cole has recorded the discovery as follows: 'The deposits contain two fossiliverous strata with a very plentiful fauna, including a hitherto unknown diminutive giraffe. Presumably they are a continuation of the Proconsul beds and they have been dated by potassium-argon to about 14 million years ago, probably late Upper Miocene and perhaps early Pliocene . . . so the new jaw may not be very much later than Proconsul.'

The human shape of the curve of the tooth row and a comparatively short face are particularly interesting features of Kenyapithecus. Leakey, when he examined the find more closely, was reminded of the upper and lower jaw-bone fragments of a primate discovered in 1935 by the Yale-Cambridge expedition to the Himalayan foothills. Because of its markedly short muzzle the discovery had been named *Ramapithecus brevirostris* (short-muzzled 'Ape of Rama' – the incarnation of the God Vishnu). One of its discoverers, G. Edward Lewis, had pointed out as early as 1938 that the ape

Jaw-bone fragment of Ramapithecus

Skull of Australopithecus, found at Sterkfontein

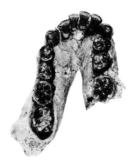

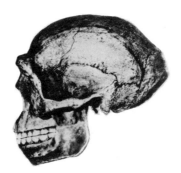

Mandible of Homo habilis, found at Olduvai

Skull of Peking man

Mandible of Heidelberg man

264

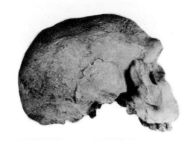
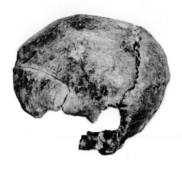

Skull of Rhodesia man   *Skull of Steinheim man (cast)*   *Skull fragment of Swanscombe man*

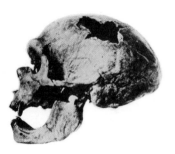

Skull of Neandertal man   *Skull of Cro-Magnon man*

they christened after a god 'could perhaps be the earliest known hominid'. But because the Siwalik deposits at the foot of the Himalayas were so very distant from the supposedly African 'Garden of Eden', and because the theory that hominids had originated in Africa was now widely accepted, the short-muzzled Siwalik primate attracted very little attention – until Leakey presented the world with its rather older African counterpart, *Kenyapithecus*.

The American palaeontologist E. L. Simons made a close study of the Kenya and Siwalik primates. As a result he not only brought them together under the name *Ramapithecus punjabicus*, but he also transferred them from hominoids to hominids. Unfortunately only very small fragments of jaws belonging to Ramapithecus have so far come to light. Analyses up till now have confirmed its status as the first demonstrably human type 'from the sub-human chapter in the history of our descent, before the decisive transformation into man', as Heberer expresses it. At a further press conference given in Kenya in January 1967 Leakey shifted the date of the oldest Ramapithecus back yet further: they constituted 'evidence that the history of the family of man now extends to more than 19 million years.'

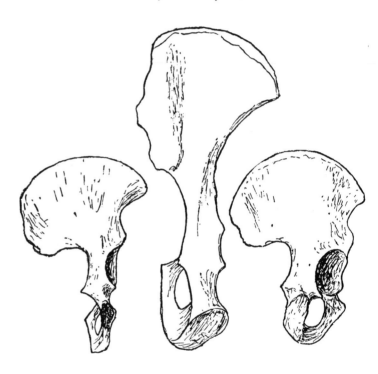

The innominate bones of the higher Primates: Plesianthropus (left), chimpanzee (centre), Bushman (right). The strong resemblance between the upper two thirds of the pelvic bones of Australopithecus and the Bushman (a modern human race), makes it seem certain that Australopithecus will have walked upright

'**Before beginning a study of man,** it is important to define what is, in fact, meant by "man". This sounds obvious and simple enough, yet when we try to arrive at such a definition we find ourselves up against considerable difficulties. Classification of a fossil skull is made after a detailed consideration of its morphological features, yet even when all the features have been studied exhaustively, it often happens that experts are in profound disagreement over the

The contours and the touch-points of (left to right) a man's hand as compared with those of the chimpanzee, the gorilla, and the orang-utan

status of the find. It must first be decided whether an individual is "man" or "ape", and, if man, to which of the groups of mankind he belongs. The line dividing man and apes is so arbitrary that considerations other than the anatomical features have to be taken into account; we are forced to fall back on defining man as a tool-maker, or as having power of speech. We depend upon his mental attributes more than the shape of his bones . . .'

In these words Sonia Cole, in her book *The Prehistory of East Africa*, outlines the age-old question which still remains unanswered despite the recent finds of sub-human apes and australopithecines: where does man begin? The pleasant illusion that it was his brain which first made a hominid into a man has been dispelled. The development of the typically human brain is apparently more recent than was first supposed. The difference in brain capacity of an ape (around 450 c.c.), of Australopithecus (400–680 c.c.), and of Asiatic early man (750–1000 c.c.) is not very great. It was not his brain, but his upright posture, allowing him to use his hands for a complex range of manipulations resulting in the use and manufacture of tools and weapons, which we now believe to have been the decisive factor in the rise of man.

'Imagine an early hominid – perhaps even a near-man,' wrote F.C. Hockett and R. Ascher in their report *The Human Revolution*, 'as he sits on the ground hitting a nut

267

In north-eastern Australia the Aborigines still shape their stone tools by direct percussion, using a hammerstone

Indirect percussion with a hammer and chisel, as practised in North America

The making of tools by pressure flaking, as practised in North America

Flaking with the aid of a chisel, as practised in north-western Australia

with a suitable stone. An attacker approaches. Our hero springs to his feet and runs away on two legs as fast as he can; but there are no trees nearby to which he can flee. However, he still holds the stone – merely because he has no immediate need of the hand. Cornered, he hits his enemy with the stone in his hand, or perhaps throws it in his direction.' This unpremeditated use of a natural aid could have marked the beginning of the manner in which hominids finally became tool-makers, and thus '*Homo faber*', cultural man.

Dr John Napier, the primatologist, sees the long-drawn-out process of the development of the use of tools and their manufacture as follows:

1. Casual use of tools
2. Intentional use of tools
3. The shaping of tools for an indefinite purpose
4. The shaping of tools for some definite purpose
5. Casual manufacture of tools
6. Cultural manufacture of tools

The first three stages of Napier's scheme do not only apply to man but to anthropoid apes as well. Whether Ramapithecus, which some experts think does not really belong to the race of hominids in the narrow sense of the term, was already a tool-maker we do not know. Heberer and other anthropologists think it possible. The intentional manufacture of tools for some future purpose has only been actually proved to apply to the australopithecines.

It is no longer possible now to distinguish sharply between animal and man – unless one groups all australopithecines together as an 'ape stage of man', making true man only begin where he has left us clear evidence of cultural and perhaps even artistic activity. But drawing a distinction of this sort would merely amount to a desperate attempt to bolster up our pride. Certainly we modern men are chiefly distinguished from all other living animals by the development of speech and our ability to think in abstract terms – but where word symbolism and the formation of abstract concepts began in fossil hominids is of course something we cannot determine.

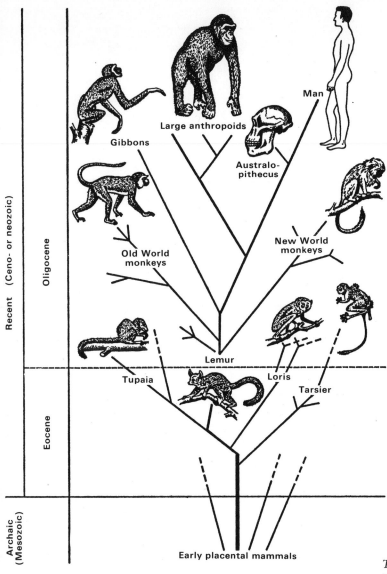

Recent (Ceno- or neozoic)

Oligocene

Eocene

Archaic (Mesozoic)

Man

Large anthropoids

Gibbons

Australo-
pithecus

New World
monkeys

Old World
monkeys

Lemur

Tupaia

Loris

Tarsier

Early placental mammals

The family tree of the Primates.
The Old World and New World
monkeys as well as the ancestors of
anthropoid apes, and thus also man,
developed out of semi-ape-like fore-
fathers during the period of transi-
tion between the Eocene and
Oligocene

269

The increase in brain capacity which is typical of man was perhaps the consequence rather than the precondition for making tools; accordingly it will have been work that provided us with the actual incentive to intellectual de-development. With his brain man can conquer the world, subdue nature and explore outer space; but the very same brain – and this is the other side of the coin – has destroyed more of his own kind in a comparatively short period of geological time than any other creature before or after him. Even in 1910, before the hunting and possibly cannibalistic instincts of Australopithecus had become manifest, William James was pessimistic enough to write: 'Our ancestors have bred in us a love of fighting which goes very deep indeed; a thousand years of peace will not be enough to eradicate it.'

Whether we shall continue to be weighed down by this legacy of the past or shall manage to rid ourselves of it depends on us alone; for it is man's great intellectual capabilities which make him the only creature able to determine his own destiny. But why is it that man has progressed so far?

One of the most eminent biologists of the day, Julian Huxley, has pointed out that all successful groups of animals were small and weak in their early stages of development. A new species cannot become established all at once; a whole series of improvements must take place during the course of its evolution, and these must somehow fuse together before the species can become a new, unified organism capable of winning its struggle for survival. How this happened to the ancestors of man, till they reached the point of becoming real men, Julian Huxley describes as follows:

'Our prehuman ape ancestors were never particularly successful or abundant. For their transformation into man a series of steps were needed. Descent from the trees; erect posture; some enlargement of brain; more carnivorous habits; the use and then the making of tools; further enlargement of brain; the discovery of fire; true speech and language; elaboration of tools and rituals. These steps took the better part of half a million years; it was not until less than a

Opposite: Man becomes ever more aggressive. This Stone Age Dane has a bone spearhead still embedded in his nose

270

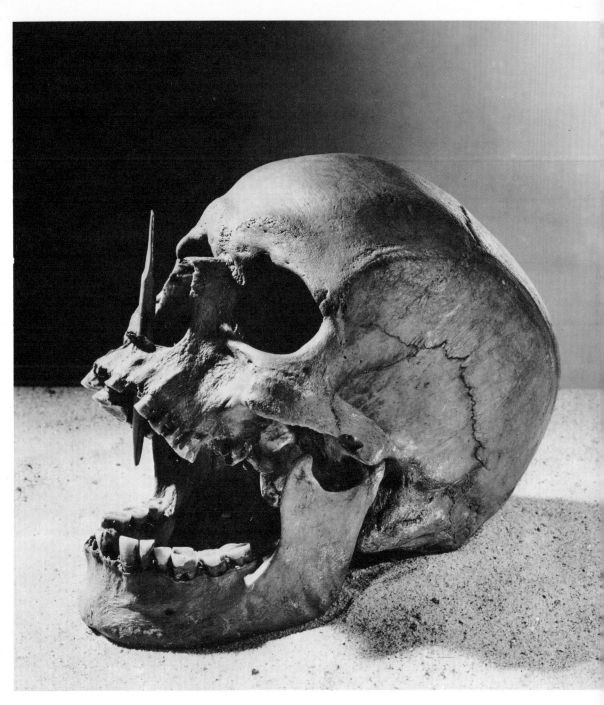

100 thousand years ago that man could begin to deserve the title of dominant type, and not till less than 10 thousand years ago that he became fully dominant.

'After man's emergence as truly man, the same sort of thing continued to happen, but with an important difference. Man's evolution is not biological but psychosocial: it operates by the mechanism of cultural tradition, which involves the cumulative self-reproduction and self-variation of mental activities and their products. Accordingly, major steps in the human phase of evolution are achieved by break-throughs to new dominant patterns of mental organization, of knowledge, ideas and beliefs – ideological instead of physiological or biological organization.'

During the later phases in our development our conscious and subconscious intellectual and spiritual impulses and attributes became more and more marked, until they formed the foundation for further evolution. From that point the essential features in the history of man have not been genetically but culturally determined. This break-through, the development of the intellect, was the result of man's dominating position in nature. However, this achievement of his also means that he is now undergoing a period of transition which can quite justifiably be called the most critical period so far in his history. The power he now has to annihilate himself by means of nuclear, chemical and radiological devices is enough to threaten the continuing existence of the human race. To this is added the spectre of over-population, which many natural scientists consider no less dangerous than the possibility of atomic war. Although our scientific and technical achievements make it possible for us to provide the prerequisites for most or even all of us to live a rich and full existence, the old limitations of politics, ideology, economic system, tradition, class and racial differences which restrict life in most of the overpopulated areas of the earth today still show no sign at all of abating. Man, so proud of his reason, has not yet managed to create a society in which all can live in peaceful coexistence, a society such as Joseph H. Muller the American geneticist so optimistically described:

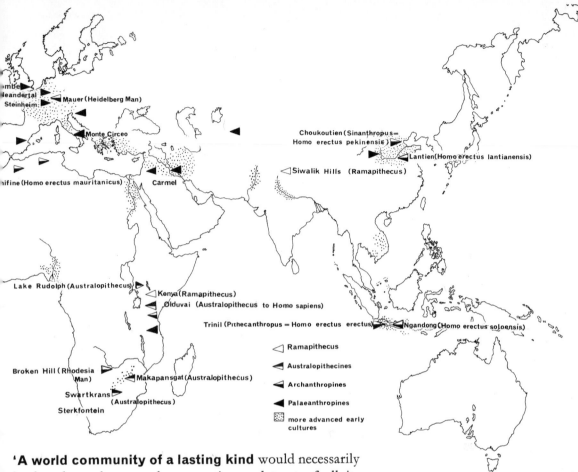

Map labels:

mbe
Neandertal
Steinheim
Mauer (Heidelberg Man)
Monte Circeo
Choukoutien (Sinanthropus = Homo erectus pekinensis)
Lantien (Homo erectus lantianensis)
Siwalik Hills (Ramapithecus)
nifine (Homo erectus mauritanicus)
Carmel
Lake Rudolph (Australopithecus)
Kenya (Ramapithecus)
Olduvai (Australopithecus to Homo sapiens)
Trinil (Pithecanthropus = Homo erectus erectus)
Ngandong (Homo erectus soloensis)
Broken Hill (Rhodesia Man)
Makapansgat (Australopithecus)
Swartkrans (Australopithecus)
Sterkfontein

◁ Ramapithecus
◀ Australopithecines
◀ Archanthropines
◀ Palaeanthropines
▨ more advanced early cultures

'**A world community of a lasting kind** would necessarily be based on the mutual cooperation and trust of all its members. It would allow greater opportunity for enlightenment, for the development of their own personality, and for a life rich in experience. It would give them wide scope in their choice of work and pleasure, but would also make them aware of the part they were playing in the conservation and progress of mankind; and it would let all men take part in the decisions reached about matters directly concerning themselves.'

Man can tame the forces of nuclear energy and use it for peaceful means. He can undertake gigantic, earth-shattering operations to transform our planet to the full advantage of

Map of the most important sites where human fossils have been found

273

all living things. And by means of progressive automation he can hand over the more tedious jobs he has to do to the machines he has invented. He can leave his own planet and venture out to other planets and satellites of our solar system. But he can just as easily destroy himself through overpopulation and the misuse of his own inventions and discoveries. True, he has made breathtaking advances in his knowledge of the world; but perhaps he has neglected to make sure he is master of his own being.

In all past ages in the history of mankind most of those who peered into the future will probably have thought that the current, familiar state of things would continue much as it was. Man stood at the centre of his world; he regarded himself as being, not evolving. But many people have not yet properly absorbed the fact that the revolution triggered off by Darwin's theory has utterly destroyed this anthropocentrism. Once the theory of evolution had become established, Muller tells us, then logically it was scarcely possible to avoid the argument that man was more likely to have created God in his own image, as Bernard Shaw had so tellingly phrased it, than the other way round. Man saw himself as a unique creature among millions of other sorts of living and extinct beings, as faced with a duty involving a tremendous degree of responsibility: he had to justify his own existence, stand on his own two feet, or confess his inadequacy.

Julian Huxley adds: 'Man . . . has been ousted from his self-imagined centrality in the universe to an infinitesimal location in a peripheral position in one of a million of galaxies. Nor, it would appear, is he likely to be unique as a sentient being. On the other hand, the evolution of mind or sentiency is an extremely rare event in the vast meaninglessness of the insentient universe, and man's particular brand of sentiency may well be unique. But in any case he is highly significant. He is a reminder of the existence, here and there, in the quantitative vastness of cosmic matter and its energy-equivalents, of a trend towards mind, with its accompaniment of quality and richness of existence; and, what is more, a proof of the importance of mind and quality in the all-embracing evolutionary process.'

His spiritual and intellectual qualities give man a chance no other sort of creature ever had: he can shape his own future, if only he will. 'Prometheus, who once stole fire, today tames the nuclear furies, scans the brain with electrodes, destroys the genes and then puts them together again,' is Joseph Muller's picture of the possibilities before us. 'Soon he will risk venturing out into the cosmos, and his creative activity beyond the earth will then begin in earnest. But the greatest of his tasks will be the creation of a more powerful Prometheus. If he ever uses this self-assumed prerogative he will find his greatest freedom.'

Here, where the ways of Cain and Prometheus part, the die is cast for the future of mankind.

The darker side of man: a skull trophy of the Dyaks of central Borneo

Epilogue

The last chapter in the discovery and study of 'near-men' and early men has by no means been written yet. In fact just as I was concluding this book a new site in East Africa became the focus of anthropological interest – the area north and east of Lake Rudolf on the boundary between Kenya and Ethiopia. British, French and North American scientists have discovered australopithecine remains at dozens of places in this area; and modern dating methods make the oldest of these fragments of jaw-bone and teeth to be about 5 million years old.

In August 1969 Richard Leakey, son of Louis and Mary Leakey, then twenty-five years of age, found a robust australopithecine skull in the dried-out bed of a stream near Lake Rudolf. British experts put its age at 2.6 million years. This has triggered off a tremendous search in the immediate area, which seems to surpass the Olduvai Gorge in its wealth of fossils. Scrapers, blades and other tools found in the basalt indicate that not only the robust P-type lived in these parts but the gracile A-type also, with his ability to use tools and weapons.

Thus at the latest 5 million years ago the transition between animal and man had already been made; the australopithecines, thanks to their technical and intellectual abilities, were able to take 'an active part in the process of their own evolution', as Heberer puts it. But perhaps even then there already existed among their own kind a creature who was the direct forebear of *Homo erectus* or even *Homo sapiens* – the 'mystery man', as Richard Leakey calls him.

For on the shores of Lake Rudolf an African participant in Leakey's expedition found fragments of a skull which turned out to be $2\frac{1}{2}$ million years old and yet surprisingly 'modern'-looking. As the custom of immediately conferring a scientific name upon each fresh fossil find has long been given up, this skull still lacks a Latin label. Richard Leakey thinks it is a 'prototype of *Homo erectus*', which, if this is so, will have developed out of Australopithecus three times as long ago as has hitherto been thought. Other anthropologists even believe there could have been a direct line of development from this 'mystery man' to *Homo sapiens*. At any rate this enigmatic prehistoric man seems to remove all trace of a boundary between the genera Australopithecus and Homo; we can no longer say where the 'ape-man', Australopithecus, ends and 'genuine man', Homo, begins.

Bibliography

Abel, Othenio, *Das Leben der Vorzeit*, Salzburg 1932.

Absolon, K., *Neue Funde fossiler Menschenskellette in Mähren*, Prague 1929.

Allison, A.C., Blood Groups and African Prehistory, in Balout, L. (ed.), 1955.

Andersson, J.G., *Children of the Yellow Earth*, London 1943.

Andrews, Roy Chapman, *Meet Your Ancestors*, New York 1945.

Arambourg, C., Contribution à l'étude géologique et paléontologique du Bassin du Lac Rudolphe et de la basse vallée de l'Omo, *Mission scientifique de l'Omo 1932–33*, Bd. 3, Paris, Mus. Hist. Nat., 1947.

Ardrey, Robert, *African Genesis*, New York 1965.

Augusta, J., and Burian, Z., *Prehistoric Man*, London 1960.

Bandi, Hans-Georg, *Die Steinzeit*, Baden-Baden 1960.

— and Maringer, Joh., *Die Kunst der Eiszeit*, Basle 1952.

Bibby, Geoffrey, *The Testimony of the Spade*, London 1957.

Bishop, W.W., The Later Tertiary and Pleistocene in Eastern Equatorial Africa (Wenner-Gren Foundation Symposium, Burg Wartenstein), in the press.

— and Posnansky, M., Pleistocene Environments and Early Man in Uganda, *Uganda Journ.*, 24, 1960.

Bolk, Ludwig, *Das Problem der Menschwerdung*, Jena 1926.

Bölsche, Wilhelm, *Die Abstammung des Menschen*, Stuttgart 1905.

— *Der Mensch der Vorzeit*, Stuttgart 1909.

Boucher de Perthes, J., *De l'Homme Antédiluvien et ses Œuvres*, Abbeville 1865.

Boule, Marcelin, and Vallois, H.V., *Fossil Men*, London 1957.

Brain, C.K., New Evidence for the Correlation of the Transvaal Ape-Man-Bearing Cave Deposits, in Clark, J.D. (ed.), *Journ. Roy. Anthrop. Inst.*, 77, 1957.

Brandt, Karl, *Uranfänge der Kunst*, Herford 1947.

Breuil, Henri, *Four Hundred Centuries of Cave Art*, Montignac 1952.

— and Koslowski, K., *Etudes de Stratigraphie paléolithique*, Paris 1932.

Broom, Robert, *Finding the Missing Link*, London 1950.

— and Schepers, G.W.M., The South African Fossil Ape-Man, the Australopithecine, *Transvaal Museum*, Pretoria, Mem. No. 2, 1946.

Buffon, Comte de, *Correspondance*, Paris 1860.

Casteret, Norbert, *Ten Years Under the Earth*, Harmondsworth 1952.

Clark, W.E. Le Gros, *Growth and Body Proportions in Relation to the Systematics of the Higher Primates*, London 1953.

— *History of the Primates*, London 1962.

— and Leakey, L.S.B., *The Miocene Hominoidea of East Africa*, London 1951.

Coates, Adrian, *Prelude to History*, London 1951.

Cole, S., *The Prehistory of East Africa*, London 1954.

— The Oldest Tool-Maker, *New Scientist*, 6, 1959.

Coppens, Y., Découverte d'un Australopithéciné dans le Villafranchien du Tchad, *C. R. Acad. Sci.*, Paris 1961.

Dart, R.A., *An Australopithecus of the Central Transvaal*, Pretoria 1948.

— The Predatory Implemental Technique of Australopithecus, *American Journal of Physical Anthropology*, 1949.

Darwin, Charles, *The Origin of Species*, latest ed., London 1950.

— *The Descent of Man*, London 1871.

— Autobiography, latest ed., London 1958.

Daudin, H., *De Linné à Lamarck*, Paris 1926.

Dobzhanski, Th., *Genetics and the Origin of Species*, N.Y. 1949.

Dubois, Eugen, *Pithecanthropus erectus, eine menschenähnliche übergangsform aus Java*, Batavia 1894.

Ecker, A., *Lorenz Oken*, Stuttgart 1880.

von Eickstedt, Egon, *Menschen und Menschendarstellungen der steinzeitlichen Höhlenkunst in Frankreich und Spanien*, 1952.

Fuhlrott, C., *Der fossile Mensch aus dem Neandertal*, Duisburg 1865.

Gehlen, A., *Der Mensch*, Bonn 1950.

Gieseler, Wilhelm, *Abstammungskunde des Menschen*, Oeringen 1936.

— *Die Fossilgeschichte des Menschen*, Stuttgart 1957.

Goethe, Joh. W., *Naturwissenschaftliche Skizzen*, Leipzig n.d.

Goethes Gespräche mit Eckermann, Leipzig n.d.

Goodall, Jane van Lawick, various articles in *National Geographical Magazine*.

Gorjanović-Kramberger, K., *Der paläolithische Mensch von Krapina*, Wiesbaden 1906.

Grahmann, Rudolf, *Urgeschichte der Menschheit*, Stuttgart 1952.

Grzimeks *Tierleben* Bd. XI, Munich 1969.

Haeckel, Ernst, *Natürliche Schöpfungsgeschichte*, Berlin 1879.

— Gemeinverständliche Vorträge und Abhandlungen, Bonn 1902.

Hagberg, Knut, *Carl Linnäus*, Hamburg 1946.

Hauser, Otto, *Der Mensch vor 100 000 Jahren*, Leipzig 1917.

— *Das Paradies der Urmenschen*, Jena 1926.

Heberer, Gerhard, *Das Präsapiensproblem*, Berlin 1950.

— *Neue Ergebnisse der menschlichen Abstammungslehre*, Göttingen 1951.

— *Die Fortschritte in der Erforschung der Phylogenie der Hominoidea*, Berlin 1952.

— *Begriff und Bedeutung der parallelen Evolution*, Freiburg 1952.

— Zur Beurteilung der Praehominiae, *Homo*, 1953.

— *Die Evolution der Organismen*, Jena 1943, new ed. 1954.

— *Homo – unsere Ab- und Zukunft*, Stuttgart 1968.

Hennig, E., *Der Werdegang des Menschengeschlechts*, Tübingen 1950.

Hoernes, D., and Menghin, O., *Urgeschichte der bildenden Kunst in Europa*, Vienna 1925.

Hooton, Ernest A., *Up from the Ape*, New York 1946.

Hrdlicka, Ales, *The Skeletal Remains of Early Man*, Washington 1930.

Huene, F. von, *Ist der Werdegang der Menschheit eine Entwicklung?* Stuttgart 1937.

Huxley, Julian, *Man in the Modern World*, London 1947.

— *Evolution*, London 1963.

Huxley, T.H., *Evidence as to Man's Place in Nature*, London 1864.

Irvine, William, *Apes, Angels and Victorians*, London 1955.

Jones, F.W., *Man's place among the mammals*, London 1929.

Kahn, Fritz, *Das Buch der Natur*, Zurich 1952.
Kälin, J., Die ältesten Menschenreste, *Historia Mundi*, Munich 1953.
Keith, A., *The Antiquity of Man*, 2nd ed. London and New York 1931.
— *A New Theory of Human Evolution*, London 1950.
Kellogg, W.N. and L., *The Ape and the Child*, New York 1933.
Klaatsch, Hermann, *Das Werden der Menschheit*, Leipzig 1936.
Koenigswald, G.H.R. von, *Das Pleistozän Javas*, 1939.
— *Search for Early Man*, New York 1947.
— *Gigantopithecus blacki*, New York 1885.
— *The Evolution of Man*, Ann Arbor, Mich. 1962.
— *Meeting Prehistoric Man*, London 1956.
—, Gentner, W., and Lippolt, H.J., Age of the Basalt Flow of Olduvai, East Africa, *Nature*, 192, 1961.
Köhter, Wolfgang, *Intelligenzprüfung an Menschenaffen*, Berlin 1921.
Kossinna, G., *Die deutsche Vorgeschichte*, Leipzig 1934.
Krause, E., *Charles Darwin*, Leipzig 1885.
Kühn, Herbert, *Die vorgeschichtliche Kunst Deutschlands*, Berlin 1935.
— *On the Track of Prehistoric Man*, London 1955.
— *Die Felsbilder Europas*, Stuttgart 1952.
— *Das Erwachen der Menschheit*, Frankfurt/Main 1954.
Lange, Fr. A., *Geschichte des Materialismus*, Leipzig 1873.
Lartet, Edouard, and Christy, Henry, *Reliquiae Aquitanicae*, Paris 1865.
Leakey, Louis S.B., *The Stone Age Cultures of Kenya Colony*, Cambridge 1931.
— The Status of the Kanam mandible and the Kanjera skulls, *Man*, 210, 1933.
— *Adam's Ancestors*, 4th ed. London 1953.
— *Stone Age Africa*, London 1936.
— Skull of Proconsul from Rusinga Island, *Nature*, 162, 1948.
— A new fossil skull from Olduvai, *Nature*, 184, 1959.
— An alternative interpretation of the supposed giant deciduous hominid tooth from Olduvai, *Nature*, 185, 1960.
— various articles in *National Geographical Magazine*.
Lee, *Memoirs of Baron Cuvier*, London 1833.
Ley, Willy, *The Lungfish, the Dodo and the Unicorn*, New York 1950.
— *Dragons in Amber*, New York 1951.
Lyell, Charles, *Geological Evidences of the Antiquity of Man*, London 1873.
— *Principles of Geology*, London 1876.
Lorenz, Friedrich, *Sieg der Verfemten*, Vienna 1952.
— *Die Entdeckung des Lebens*, Vienna 1949.
Lorenz, Konrad, *On Aggression*, London 1966.
MacCurdy, G.G., *Early Man*, London 1937.
Mansfield, J. Carroll, *Dawn of Creation*, London 1952.
Menghin, Oswald, *Weltgeschichte der Steinzeit*, Vienna 1931.
Montelius, O., *The Civilisation of Sweden . . .*, London 1888.
Moore, Ruth, *Man, Time and Fossils*, London 1962.
Morris, Desmond, *The Naked Ape*, London 1967.
Movius, H.L., *The Lower Palaeolithic Cultures of Southern and Eastern Asia*, Philadelphia 1949.
Napier, J., and Weiner, J. S., Olduvai Gorge and Human Origins, *Antiquity*, XXVI, March 1962.
Nordenskjöld, Erik, *Die Geschichte der Biologie*, Jena 1928.
Oakley, Kenneth P., *Frameworks for Dating Fossil Man*, London 1964.

— in Clark, J.D., New Studies on Rhodesian Man, *Journ. Roy. Anthrop. Inst.*, 77, 1957.
— *Man the Tool-Maker*, 5th ed., London 1967.
Obermaier, Hugo, *Der Mensch der Vorzeit*, Berlin 1922.
— *Urgeschichte der Menschheit*, Freiburg 1931.
Portmann, Adolf, *Zoologie und das neue Bild des Menschen*, Hamburg 1956.
— *Der Pfeil des Humanen*, Munich 1960.
Ranke, Johannes, *Der Mensch*, Leipzig 1922.
Reck, Hans, *Die Schlucht des Urmenschen*, Leipzig 1951.
Rensch, Bernard (ed.), *Handgebrauch und Verständigung bei Affen und Frühmenschen*, Berne-Stuttgart 1968.
Reynolds, Frankie and Vernon, various articles in *National Geographical Magazine*.
Robinson, J.T., *The Australopithecines and their Evolutionary Significance*, London 1952.
— *Meganthropus, Australopithecines and Hominids*, Pretoria 1953.
— *The Nature of Telanthropus capensis*, London 1953.
Roslansky, John D., *The Human Mind*, Amsterdam 1967.
Rostand, Jean, *Die Biologie und der Mensch der Zukunft*, Darmstadt 1953.
Schaaffhausen, Hermann, *Der Neandertaler Fund*, Bonn 1888.
Schoetensack, O., *Der Unterkiefer des Homo heidelbergensis*, Leipzig 1908.
Schwalbe, Gustav, *Der Neandertalschädel*, Bonn 1901.
Schwantes, Gustav, *Deutschlands Urgeschichte*, Stuttgart 1952.
Simon, W., *Die Uhren der Erde*, Clausthal 1953.
Simpson, George G., *The Meaning of Evolution*, London 1950.
Soergal, W., *Die Jagd der Vorzeit*, Jena 1922.
— *Das Eiszeitalter*, Jena 1938.
Sterne, Carus, *Werden und Vergehen*, Berlin 1905.
Stresemann, Erwin, *Die Entwicklung der Ornithologie*. Berlin 1951.
Taylor, Gordon Rattray, *The Science of Life*, London 1963.
Teilhard de Chardin and Weng Chung-pei, *The Lithic Industry of the Sinanthropus Deposits in Chou-kou-tien*, Bull. Geol. Soc. China, XI, 1932.
Thesing Curt, *Geheimnisse des Lebens*, Munich 1947.
Unger, Hellmuth, *Virchow*, Hamburg 1953.
Vallery-Radot, R., *Louis Pasteur*, Paris 1900.
Weidenreich, F., *Giant Early Men from Java and South China*, Anthr. Papers Amer. Mus. Nat. Hist. 40, pt. 1, 1945.
— *Morphology of Solo Man*, New York 1951.
— *Apes, Giants and Man*, Chicago 1946.
Weiner, J.S., The Pattern of Evolutionary Development of the Genus Homo, *S. Afr. Journ. Med. Sci.*, 23, 1958.
—, Oakley, K.P., and Clark, W.E. Le Gros, *The Solution of the Piltdown Problem*, Bull. Brit. Mus. (Nat. Hist.) Geology, vol. 2, no. 3, 1953.
Weinert, Hans, *Pithecanthropus erectus*, Stuttgart 1928.
— *Ursprung der Menschheit*, Stuttgart 1932.
— *Entstehung der Menschenrassen*, Stuttgart 1938.
— *Der geistige Aufstieg der Menschheit*, Stuttgart 1940.
— *Stammesentwicklung der Menschheit*, Brunswick 1951.
Yerkes, Robert, and Learned, Blanche, *Chimpanzee Intelligence*, Baltimore 1925.
Yerkes, Robert and Ada, *The Great Apes*, New Haven 1929.
Zeuner, F.E., Climate and Early Man in Kenya, *Man*, 14, 1948.
— *Dating the Past*, London 1952.
Zotz, Lothar, *Vormenschen, Urmenschen und Menschen*, Stuttgart 1948.

List of Illustrations

Numerals refer to the number of the page

8 The Creation of Man, from J.J. Scheuchzer *Physica Sacra*, 1731.

12 Statue found in the sea off Antikythera. Bronze, *c.* 340 BC. Athens, National Museum. Photo Hirmer

15 Silver gilt bowl of Phoenician design found in the Bernardini tomb at Palestrina (Praeneste). 7th century BC. Museo Preistorico L. Pigorini, Rome. Photo Soprintendenza alle Antichità, Rome

Detail of drawing of the Praeneste bowl, from O. Montelius *La Civilisation Primitive en Italie*, 1904

17 Tondo of Greek kylix. By courtesy of the Trustees of the British Museum, London

18 Hanuman, the monkey-god, from a Ceylonese shrine. Bronze, 11th century or later. London, Victoria and Albert Museum

19 Drawing of a relief of Khnemu the Potter modelling the first human being out of clay

20 Drawing: comparative anatomical studies by Leonardo da Vinci. Windsor Castle. By gracious permission of Her Majesty the Queen

21 Engraved title-page of A. Vesalius *De humani corporis fabrica*, 1543

Drawing from A. Vesalius *De humani corporis fabrica*, 1543

22 Julien Offroy de la Mettrie (1709–51): engraving by G.F. Schmidt. Paris Bibliothèque Nationale. Photo F. Foliot

23 Mammoth teeth, from M. Merian, *Theatrum Europaeum*, 1645–47

24 Giants, from A. Kircher *Mundus Subterraneus*, 1878

The giant Vienna bone, from O. Abel *Paleabiologie*, 1912

26 The Würzburg fake fossils. Würzburg, Mainfränkisches Museum. Photo Zwicker

27 Fossil skeleton discovered by Scheuchzer, from J. Scheuchzer *Physica Sacra*, 1731

29 Portrait of Scheuchzer, from J.J. Scheuchzer *Physica Sacra*, 1731

Title page and frontispiece from J.J. Scheuchzer *Museum Diluvianum*, 1716

30 Fossil bones, from J.K. Esper *Ausführliche Nachrichten*, 1774

31 Portrait of Karl von Linné (1707–78) by H. Kingsbury after M. Hoffman, 1737. Gripsholm Castle. Photo Swedish National Portrait Gallery

32 Four candidates for the 'missing link', from C.E. Hoppius *Anthropomorpha*, 1760

33 'The Anatomy lesson of Dr Tulp' by Rembrandt, 1632. The Hague, Mauritshuis

Tyson's dissection of a chimpanzee, from E. Tyson *Orang-Outang, sive Homo Sylvestris*, 1699

34 Esau, from J.J. Scheuchzer *Physica Sacra*, 1731

35 Anatomical drawings of an Orangutan from P. Camper *Oeuvres*, 1803

36 Goethe in Italy by Tischbein, 1787. Städelsches Kunstinstitut, Frankfurt

38 Engraving of Buffon examining the mammalian reproductive organs of a bitch, from G.L. Leclerc, Comte de Buffon *Histoire Naturelle*, 1749

39 Imaginative reconstruction by E. Fichel of a meeting of naturalists at the Jardin du Roi, with Buffon in the centre, 1872. Photo F. Foliot

40 Portrait of Lamarck by Boilly. Paris, Musée de l'Homme

41 View of the Jardin des Plantes, Paris *c.* 1820. Paris, Musée de l'Homme

42 Engraved portrait of G. Cuvier by Migu, after Vincent. Paris, Musée de l'Homme

Reconstruction of a megatherium, from G. Cuvier *Recherches sur les Ossemens Fossiles*, 1812

43 Progress in axes, from Kenneth P. Oakley *Man the Tool Maker*, by permission of the Trustees of the British Museum (Natural History)

44 C.J. Thomsen showing members of the public around the Old Nordic Museum, Copenhagen. Drawing by Magnus Petersen, 1846

45 Polished stone battle-axes. Copenhagen, National Museum

Double-headed flint axe. Copenhagen, National Museum

46 Flint dagger from Hindsgavl. Late Neolithic. Copenhagen, National Museum

Shell necklace from Ofnet, Bavaria, from R.R. Schmidt *Die Diluviale Vorzeit in Deutschland*

47 Late Stone Age settlement, reconstruction painting by J. Augusta and Z. Burian. By courtesy of the State Pedagogical Publishing House, Prague

48 Bust of P. Schmerling. Brussels, Palais des Académies. Photo ACL

Drawing of skulls, from P. Schmerling *Les Ossemens Fossiles*, 1833

49 Section of the Welsh cave known as 'Goat's Hole', from W. Buckland *Reliquae Diluvianae*, 1823

50 Buckland lecturing on geology in the Ashmolean Museum, Oxford, drawing by N. Whitlock. Oxford, Museum of the History of Science

William Pengelly, from H. Pengelly *William Pengelly*, 1897

51 Portrait of Boucher de Perthes. Paris, Musée de l'Homme

52 Some of Boucher de Perthes' once-disputed flint implements, from his *Antiquités Celtiques*, 1857

53 Sir Charles Lyell, drawing by George Richmond. On loan to the Scottish National Portrait Gallery from the National Portrait Gallery, London

54 Glaciers near Monte Rosa, from L. Agassiz *Etudes sur les Glaciers*, 1840

56 Skull-cap of a Neandertaler, found 1856. Bonn, Landesmuseum

57 Portrait of Rudolf Virchow. Photo CIBA, Basle

Head of Neanderthal man, reconstruction by Maurice Wilson (cast). By permission of the Trustees of the British Museum (Natural History)

59 Portrait of Charles Darwin at the age of thirty, by George Richmond. Darwin Museum, Down House, Kent. By courtesy of the President, Royal College of Surgeons, England

61 Portrait by Francis Lane of Robert Fitzroy (1805–65), as Vice Admiral. Greenwich, Royal Naval College, Hospital Collection

Marine iguanas of the Galapagos Islands. Photo Rudolf Freund

62 Natives of Tierra del Fuego, showing their appearance during and after their stay in England in 1832. Drawing by Captain Fitzroy in his *Narrative of the Surveying Voyages of HMS Adventure and Beagle*, 1839

64 Charles Darwin's study at Down House, etching by A. Haig, 1882. Darwin Museum, Down House, Kent. Reproduced by courtesy of the President, Royal College of Surgeons, England

65 Alfred Russel Wallace in the jungle of Borneo. Moscow, Museum Darwinianum. Photo by courtesy of Lady Darwin

67 Caricature of Charles Darwin from *The Hornet*, 1871. Galton Laboratory of University College, London, property of Professor E. S. Pearson

68 First page of Darwin's outline for the *Origin of Species*. This is a draft of an earlier version of the work from which the shorter published edition (1859) was abstracted. Cambridge University Library

69 Cartoon of Bishop Wilberforce from *Vanity Fair*, 1860

70 Drawing of Thomas Henry Huxley by his daughter, Marian Collier. London, National Portrait Gallery

71 Huxley's sketch of Eohomo and Eohippus, 1876

72 Portrait of Charles Darwin in old age by J. Collier. London, National Portrait Gallery. Photo Royal College of Surgeons, England

73 Drawing of Karl Vogt. Karl Sudhoff Institut, Leipzig

74 Baby lying on its stomach. Photo Comet, Zurich

75 Baby gorilla crawling. Photo Elsbeth Siegrist, Basle

76 Cartoon from *Die Brücke*, Munich

77 Haeckel at the microscope, Berlin 1858. From *Ernst Haeckel im Bilde*, 1914

78 Evolutionary tree, from Ernst Haeckel *Anthropogenie*, 1874

81 Comparison of embryos, from Ernst Haeckel *Anthropogenie* 1874

82 Ernst Haeckel (1834-1919). Photo National Library of Medicine, Washington

83 Painting of *Pithecanthropus alalus* by Gabriel Max, from Ernst Haeckel *Schöpfungsgeschichte*, 1911

84 Bison and horse. Cave painting, Lascaux (Dordogne)

87 Map of Europe and Asia during the Ice Age. After E. von Eickstedt

88 The major glaciations during the Pleistocene epoch. After Pilbeam

89 Engraved mammoth tusk, from E. Lartet and H. Christy *Reliquiae Aquitanicae*, 1865-75

Carvings on bone and ivory, from E. Lartet and H. Christy *Reliquiae Aquitanicae*, 1865-75

90 Cro-Magnon skull, frontal and lateral view. Paris, Musée de l'Homme

91 Encampment of late Palaeolithic mammoth hunters. From J. Augusta and Z. Burian *Prehistoric Man*. By courtesy of the State Pedagogical Publishing House, Prague

93 Late Palaeolithic cave artists painting pictures featuring arrows. From J. Augusta and Z. Burian *Prehistoric Man*. By courtesy of the State Pedagogical Publishing House, Prague

94/95 Table: Classification of hand-axes in Europe. After Mortillet

98 Altamira: painted ceiling. Group of bison surrounded by a hind, a boar, two horses and a second boar

99 Drawing by Breuil of paintings on main ceiling at Altamira, from E. Cartailhac and H. Breuil *La Caverne d'Altamira*, 1906

100 Comte Bégouen, J. Bégouen, M. Bégouen, Abbé Breuil, Louis Bégouen and Emile Cartailhac at the entrance to the cave at Tuc d'Audoubert, 12 October, 1912, 48 hours after discovering the clay bison. Photo Musée de l'Homme, Paris

101 Carvings in reindeer antler. Length 14 cm. St Germain-en Laye Musée des Antiquités Nationales. Photo Archives Photographiques

Engraving on a pebble of a standing bison from Laugerie Basse, Dordogne. Length 10 cm. Paris, Musée de l'Homme. From a cast in the Institute of Archaeology, London. Photo Peter Clayton

Reindeer carved on antler from the Kesslerloch, Thayingen. Drawing from *Mitteilungen der Antiquarischen Gesellschaft*, Band 18, Heft 15

102 Bull and horse: reproduction in line of cave painting at La Mouthe (Dordogne)

104 Horse engraved on rock surface at Les Combarelles (Dordogne). Length 68 cm. Photo Archives Photographiques

105 Engraved horse's head from Font-de-Gaume (Dordogne)

Two horses' heads carved out of antler. From Mas d'Azil (Ariège)

106 Reproduction of drawing by Breuil of the 'little mammoth' at Font-de-Gaume from L. Capitan, H. Breuil and D. Peyrony *La Caverne de Font-de-Gaume*, 1910

Polychrome bison from Font-de-Gaume, as sketched by Breuil, from L. Capitan, H. Breuil and D. Peyrony *La Caverne de Font-de-Gaume*, 1910

107 Finger-drawing of woman and mammoth on the ceiling of the big chamber of the cave at Pech-Merle

108 Rock painting of horses and stencilled hands. Cave of Pech-Merle. Width of panel 340 cm.

109 Rhinoceros frieze at Rouffignac

110 Breuil's reconstruction drawing of 'The Sorcerer' in Les Trois Frères cave (Ariège)

'The Sorcerer'. Engraving, painted black, in Les Trois Frères cave

113 Feline carved in reindeer antler from Isturitz (Basses Pyrenées). Length 9.5 cm. St Germain-en-Laye, Musée des Antiquités Nationales. Photo Archives Photographiques

A group of bâtons from Le Placard (Charente) ornamented with animal heads. St Germain-en-Laye, Musée des Antiquités Nationales. Photo Archives Photographique

115 Lascaux, general view of main hall. Photo Laborie. Bergerac

116 Lascaux, rotunda with painting of a variety of animals

117 'Shaft of the Dead Man', Lascaux

118 Willendorf Venus. Figure in carboniferous limestone. Height 11 cm. Vienna, Naturhistorisches Museum

119 Ivory female figurine from Lespugue. Paris, Musée de l'Homme. Photo R. Pasquino

120 Female figurine in baked clay from Dolní Vestonice (Central Moravia). Brno Museum. Photo J. Kleibl

Fragments of male figurine, carved from mammoth tusk, from Brno (Central Moravia). Gravettian culture. Photo J. Kleibl

121 Ivory head, from Dolní Vestonice. Height 4.8 cm. Brno Museum. Photo J. Kleibl

Ivory head, from Brassempouy (Landes). Height 3.65 cm. St Germain-en-Laye, Musée des Antiquités Nationales. Photo Archives Photographiques

122 Reconstruction drawing of the 'dancing women' of Cogul, Spain

123 Painting in black of a group of archers from Cueva del Civil, Valltorta gorge, Castellón. Reconstruction by Douglas Mazonowicz

125 Rock engraving of human figures at Monte Pellegrino, Sicily. Height of panel about 100 cm. Photo Soprintendenza alle Antichitá, Palermo

126 Seated figure with fan-shaped head, in red paint, at Levanzo, Sicily. Height 30 cm. Photo F. Teegen

Henri Breuil examining the painting of the 'White Lady' at the University of Johannesburg. Photo Musée de l'Homme, Paris

127 Skeletons found at Grimaldi, near Monte Carlo. Monaco, Musée d'Anthropologie. Photo by permission of the Trustees of the British Museum (Natural History)

128 The Asselar skull, from *Archives de l'Institut de Paléontologie humaine*, Mémoire no. 9

130 Drawing of a Mesolithic burial at Tèviec. Philip R. Ward

Base of the Monte Circeo skull, from a cast in the British Museum (Natural History). By permission of the Trustees of the British Museum (Natural History)

131 Cannibals of Krapina (Moravia). From J. Augusta and Z. Burian *Prehistoric Man*, by courtesy of the State Pedagogical Publishing House, Prague

133 Otto Hauser at Le Moustier. Photo by courtesy of the Institut d'Anthropologie de l'Université de Genève

Skull found at Le Moustier, August 12, 1908. Photo by courtesy of the Institut d'Anthropologie de l'Université de Genève

134 Aurignacian man found in his original position in an Old Stone Age grave. Museum für Vor- und Frühgeschichte der Staatlichen Museen, Preussischer Kulturbesitz, Berlin. Foto Croon

135 Neandertal skull, La Chapelle-aux-Saints. Paris, Musée de l'Homme

138/39 Reconstructions illustrating the evolution of man. Left to right: Proconsul, Australopithecus, Peking man, Neandertal man, Mount Carmel man, Cro-Magnon man. Drawing by Maurice Wilson

141 The Steinheim skull, cast. Lateral view

142 Reconstruction drawing of Swanscombe man, by Maurice Wilson. By permission of the Trustees of the British Museum (Natural History)

143 The Swanscombe skull. By permission of the Trustees of the British Museum (Natural History)

Stone hand-axe found at Swanscombe, Kent. Two-thirds natural size. By permission of the Trustees of the British Museum (Natural History). Photo by courtesy of Dr Kenneth P. Oakley

144 Yew spear found at Clacton. By permission of the Trustees of the British Museum (Natural History)

146 Reconstruction painting of the head of Pithecanthropus from J. Augusta and Z. Burian *Prehistoric Man*. By courtesy of the State Pedagogical Publishing House, Prague

149 Discussing the Piltdown skull. Left to right: T.D. Barlow, Professor Elliott Smith, Professor A.S. Underwood, Professor A. Keith, Charles Dawson, Dr Smith Woodward, W.P. Pycraft, Sir E. Ray

Lancaster. After a painting at the Geological Society. By permission of the Trustees of the British Museum (Natural History)

151 'Searching for the Piltdown Man'. Tableau postcard showing Charles Dawson (left) and Sir Arthur Smith-Woodward searching for the Piltdown skull

152 Two reconstruction busts of Piltdown man, made by Maurice Wilson for the Festival of Britain 1951. Photo Associated Press

153 Two versions of the Piltdown skull. Left, Sir Arthur Smith-Woodward's reconstruction drawing, right Sir Arthur Keith's version from the cover of his book *Antiquity of Man*, 1915

154 Kenneth Oakley at the Natural History Museum, London, discussing with L.E. Parsons (right) how the ape-like jaw-bone found at Piltdown could be sampled with least risk of damage. Photo Neave Parker

155 Skull of Rhodesia man, lateral view. By permission of the Trustees of the British Museum (Natural History)

Bust reconstruction model by Maurice Wilson of Rhodesia man. By permission of the Trustees of the British Museum (Natural History)

156 Tools of Rhodesia man; bone point or awl; bone gouge; spherical missile made of granite. From Kenneth P. Oakley *Man the Toolmaker* by permission of the Trustees of the British Museum (Natural History)

157 University of Cambridge, new radiocarbon dating laboratory. Photo Radio-Carbon Laboratory, Cambridge

158 Professor W.F. Libby holding a sample of reed matting and rope in front of the combustion system of the radiocarbon dating apparatus at the University of California, Los Angeles. Photo by courtesy of University of California, Los Angeles

159 Dating by the potassium-argon method. Photograph by Dean Conger, copyright National Geographic Society

160 Memorial to Heidelberg man at Mauer. Photo Bildstelle, Landratsamt Heidelberg

162 Heidelberg mandible. Photo Bildstelle, Landratsamt Heidelberg

163 Excavations at Vértesszöllös. Photo by courtesy of Dr László Vértes

164 The Vértesszöllös skull. By permission of the Trustees of the British Museum (Natural History)

165 The Trinil memorial stone

166 Sectional elevation of the site at Trinil, drawing by Eugène Dubois

167 The Trinil skull-cap (Pithecanthropus I), left lateral view. Photo by courtesy of Professor J.S. Weiner and Dr D.A. Hooiser

168 Drawing of Dubois' first reconstruction of the Pithecanthropus skull, 1896

169 *Pithecanthropus modjokertensis* (Pithecanthropus IV) Lower Pleistocene, Sangiran, central Java. Recent reconstruction model by von Koenigswald. Photo by courtesy of Professor G.H.R. von Koenigswald

170 Bust reconstruction model by Maurice Wilson of Java man. By permission of the Trustees of the British Museum (Natural History)

172 Reconstruction painting by Maurice Wilson of *Pithecanthropus erectus* (Java man). By permission of the Trustees of the British Museum (Natural History)

174/5 Skull fragments and crania of Solo man from Ngandong. Photos by courtesy of Professor G.H.R. von Koenigswald

176 Modjokerto infant calvaria. Photo by courtesy of Dr Michael Day

177 Excavation site of Pithecanthropus III, Sangiran, central Java, 1938. Photo by courtesy of Professor G.H.R. von Koenigswald

The first Pithecanthropus mandible from Sangiran. Photo by courtesy of Professor G.H.R. von Koenigswald

178 Reconstruction painting of Pithecanthropus I beside the Solo River, Java. From J. Augusta and Z. Burian *Prehistoric Man*, by courtesy of the State Pedagogical Publishing House, Prague

179 Stone implements from the Trinil level at Sangiran

180 Teilhard de Chardin on a very old bridge on the Imperial Road near Sienshien c. 1924. Photo by courtesy of the Fondation Teilhard de Chardin, Paris

182 Model of a prominent 19th-century Canton pharmacy c. 1880. London, Wellcome Institute. Reproduced by courtesy of the Trustees

185 Chou-kou-tien 1929. Left to right: Pei Weng-chung, C.C. Young (formerly Yang), two students, Teilhard de Chardin, Davidson Black, and George B. Barbour. Photo by courtesy of the Fondation Teilhard de Chardin, Paris

Drawing of the human-looking molar found by Birgir Bohlin at Chou-kou-tien on October 16, 1927

186 Excavations in progress at Chou-kou-tien. By permission of the Trustees of the British Museum (Natural History)

187 Bust reconstruction model by Maurice Wilson of Peking man. By permission of the Trustees of the British Museum (Natural History)

188 Reconstruction painting by Maurice Wilson of the Chou-kou-tien site with *Pithecanthropus pekinensis*. By permission of the Trustees of the British Museum (Natural History)

Skull of Peking man (cast). By permission of the Trustees of the British Museum (Natural History)

189 Drawing of thigh bones of Peking man found at Chou-kou-tien

Below: bone implements fashioned by our ancestors in China

190 Table of finds relating to early man (*Homo erectus*) (after G. Heberer and G. Kurth) from G. Glowatzki *Tausend Jahre wie ein Hauch*, Franckh'sche Verlagshandlung, Stuttgart

192 Flake-tool of chert, found at Chou-kou-tien. From Kenneth P. Oakley *Man the Toolmaker*. By permission of the Trustees of the British Museum (Natural History)

193 Henri Breuil's first visit to Chou-kou-tien in August 1931. Left to right: Pei, Wong and Breuil. Photo Musée de l'Homme, Paris

194 Fragment of the lower jaw-bone of *Meganthropus paleojavanicus*. Photo by courtesy of Professor G.H.R. von Koenigswald

195 View from above of the jaw-bone of *Meganthropus paleojavanicus*. Photo by courtesy of Professor G.H.R. von Koenigswald

196 Comparison of jaw-bones: (left to right) modern man, meganthropus (reconstructed by Professor Wiedenreich) and gorilla. Courtesy of the American Museum of Natural History

199 Ternifine mandible, right lateral view. Photo by courtesy of Professor C. Arambourg

Hand-axe, made of petrified lava, found in Bed II at Olduvai

200 Skull fragment of East African early man, *Homo erectus leakeyi*, from Bed II at Olduvai

202 'Primeval man ponders the World': caricature by Gourmelin, from *Bizarre*

205 Possible interpretation of hominid evolution. After Desmond Clark

206 Locations of important sites in South Africa where the remains of austrolopithecines were found

207 Raymond Dart displays the child's skull found at Taung, Botswana, 1924

210 The Taung skull, right lateral view. By courtesy of Professor P.V. Tobias. Photo A. R. Hughes

211 Taung skull, frontal view. By courtesy of Professor P.V. Tobias. Photo A.R. Hughes

Reconstruction drawing by Broom of the head of the Taung ape-man child, *Australopithecus africanus Dart*. From R. Broom *Finding the Missing Link*, C.A. Watts & Co. Ltd

212 Robert Broom in South Africa. By permission of the Trustees of the British Museum (Natural History)

214 Hypothetical sketch by Robert Broom of 'Dart's child'. From C. Andrews *Meet your Ancestors*, John Long Ltd

Comparison of the skulls of a chimpanzee and *Australopithecus transvaalensis*. Drawn by M. Maitland Howard after Le Gros Clark

215 Robert Broom and John Talbot Robinson at Sterkfontein. Photo Africana Museum, Johannesburg

216 Broom (right) and Robinson examining fossils at Sterkfontein. Photo Africana Museum, Johannesburg

219 Reconstruction drawings by Broom of the heads of the female Sterkfontein ape-man, *Plesianthropus transvaalensis* and *Paranthropus robustus Broom*. From R. Broom *Finding the Missing Link*, C.A. Watts and Co. Ltd

220 Reconstruction model by Maurice Wilson of *Plesianthropus transvaalensis*, based on the female skeleton from Sterkfontein, Transvaal. By permission of the Trustees of the British Museum (Natural History)

221 Right lateral view of skull of female Australopithecus (*Plesianthropus transvaalensis*). By courtesy of Professor J.T. Robinson

222 Australopithecines in their South African habitat. From J. Augusta and Z. Burian *Prehistoric Man*. By courtesy of the State Pedagogical Publishing House, Prague

224 Raymond Dart showing how the shoulder blade of an ox could be used to kill prey. Photo Johannesburg Star

Bone tools used by Australopithecus in South Africa. By courtesy of Professor P.V. Tobias. Photo A.R. Hughes

225 An ungulate thigh-bone used as a weapon by australopithecines

227 View of Makapansgat from the mouth of the Makapan cave, northern Transvaal. Photo by courtesy of the Trustees of the British Museum (Natural History)

228 A mandible from Swartkrans, SK 15 (Telanthropus I), occlusal view. By courtesy of Professor J.T. Robinson

231 Louis Leakey comparing the reconstructed skull of Zinjanthropus with the smaller skull of a modern chimpanzee. The bone connecting the skull with the neck indicates that Zinjanthropus walked upright like a man. Photo USIS

233 Sketch map of the Olduvai Gorge on the edge of the Serengeti plain, and geological section through the face and the Balbal depression. From G. Heberer *Homo – Unsere Ab- und Zukunft*, Deutsche Verlagsanstalt, Stuttgart

234 Reconstruction of skull of Zinjanthropus, frontal view. By permission of the Trustees of the British Museum (Natural History)

235 Palate of *Zinjanthropus boisei* compared with those of an Australian aborigine modern man and gorilla. Photo Jen and Des Bartlett and Bruce Coleman Ltd

236 The two halves of the Zinjanthropus palate being uncovered in Olduvai Gorge. Dental pick held by Dr Leakey, July 1959. Photo Jen and Des Bartlett and Bruce Coleman Ltd

Louis and Mary Leakey at work in camp at Olduvai Gorge having just joined the two halves of the Zinjanthropus palate. Photo Jen and Des Bartlett and Bruce Coleman Ltd

238 Kenneth Oakley at Sterkfontein, 1953. Photo by C.K. Brain, by courtesy of Dr Kenneth P. Oakley

239 Skull of Zinjanthropus (right) is compared with the skull of an Australian aborigine. Photo Jen and Des Bartlett and Bruce Coleman Ltd

240 Pebble tools as used by the australopithecines (Zinjanthropus) to skin animals. Photo by Kathleen Revis, copyright National Geographic Society

241 Reconstruction painting by Maurice Wilson of Australopithecus. By courtesy of the Trustees of the British Museum (Natural History)

242 Reconstruction in half-section of Australopithecus with and without the sagittal crest. From G. Heberer *Homo – Unsere Ab- und Zukunft*. Deutsche Verlagsanstalt, Stuttgart

244 Olduvai hand (*Homo habilis*). By courtesy of Dr L.S.B. Leakey

245 Olduvai foot (*Homo habilis*). By courtesy of Dr L. S. B. Leakey

246 Reconstruction drawing by Barry Driscoll of *Homo habilis*. Photo Sunday Times

248 Comparison of the chromosome series of the higher Primates, from G. Heberer *Homo – Unsere Ab- und Zukunft*, Deutsche Verlagsanstalt, Stuttgart

249 Comparison of the facial muscles of man and anthropoid ape. Reproduced from J. Biegert 'Vom Ursprung und Werden des Menschen' in *Schweizer Illustrierte*, No. 45, 1965

250 The male orang-utan 'Alexander' with one of his paintings. Fox Photos

Brush painting by the young male chimpanzee 'Congo', showing typical fan pattern. From Desmond Morris *The Biology of Art*

252 The five-year-old chimpanzee 'Fifi' uses a tool she herself made by stripping down a grass stalk. Photo Baron Hugo van Lawick, copyright National Geographical Society

253 Jane van Lawick-Goodall with the 11-month-old wild chimpanzee 'Flint'. Photo Baron Hugo van Lawick, copyright National Geographic Society

254 The female chimpanzee 'Julia' deftly fits a small key into a padlock she had not encountered before. After B. Rensch and J. Döhl, from *Handgebrauch und Verständigung bei Affen und Frühmenschen*, Verlag Hans Huber

255 The female chimpanzee 'Julia' shows her skills during intelligence tests. After B. Rensch and J. Döhl, from *Handgebrauch und Verständigung bei Affen und Frühmenschen*, Verlag Hans Huber

256 A forest-dwelling chimpanzee entering a plantation. Photo by courtesy of Dr A. Kortland

A chimpanzee captured in the savannah, attacking a dummy leopard. Photo by courtesy of Dr A. Kortland

257 A forest-dwelling chimpanzee preparing to attack a dummy leopard. By courtesy of Dr J. C. J. van Zon

259 Louis Leakey at work on Rusinga Island. Photo by courtesy of Dr Kenneth P. Oakley

260 Proconsul skull, frontal and lateral view, and reconstruction in section. By permission of the Trustees of the British Museum (Natural History)

261 Skeleton of *Oreopithecus bamboli* in a coal seam. Basle, Naturhistorisches Museum

263 Reconstruction drawing by Maurice Wilson of Oreopithecus. By permission of the Trustees of the British Museum (Natural History)

264 Ramapithecus jaw-bone fragment. Photo Elwyn Simons

Frontal view of Australopithecus skull from Sterkfontein. By courtesy of Professor J. T. Robinson

Mandible of type specimen of *Homo habilis* (Olduvai hominid 7). By courtesy of Dr L. S. B. Leakey

Left lateral view of skull of *Pithecanthropus pekinensis*. By permission of the Trustees of the British Museum (Natural History)

Heidelberg mandible. Photo Bildstelle, Landratsamt Heidelberg

265 Skull of Rhodesia man, frontal view. By courtesy of the Trustees of the British Museum (Natural History)

Cast of the Steinheim calvarium, right lateral view

Fragment of the Swanscombe skull. By courtesy of the Trustees of the British Museum (Natural History)

Left lateral view of Neandertal skull found at St Chapelle-aux-Saints

Frontal view of Cro-Magnon skull. Paris, Musée de l'Homme

266 Innominate bones of the higher Primates, drawn by Robert Broom. From R. Broom *Finding the Missing Link*, C. A. Watts & Co. Ltd

267 The contours and the touch-points of (left to right) a man's hand compared with those of the chimpanzee, the gorilla, and the orang-utan. From J. Biegert 'Vom Ursprung und Werden des Menschen' in *Schweizer Illustrierte*, No. 45, 1965

268 Methods of flaking stone: (from top to bottom) direct percussion with hammerstone, N.E. Australia; indirect percussion, N. America (after W. H. Holmes); pressure flaking, N. America (after W. H. Holmes); pressure flaking, N.W. Australia (after D. S. Davidson). From Kenneth P. Oakley *Man the Toolmaker*, by permission of the Trustees of the British Museum (Natural History)

269 Genealogical tree of the Primates. From G. Glowatzki *Tausend Jahre wie ein Hauch*, Franckh'sche Verlagshandlung, Stuttgart

271 Skull of a man with a spearhead embedded in his nose, from Pormose near Naestved, South Zealand. Copenhagen, National Museum

273 Map of the most important sites that have yielded human fossils

275 Skull trophy of the Dyaks of central Borneo. Photo by courtesy of Professor G. H. R. von Koenigswald

Colour Plates

Opposite page

48 Neandertal skull found at La Chapelle-aux-Saints. Photo by courtesy of Dr M. H. Day

Neandertal encampment. From J. Augusta and Z. Burian *Prehistoric Man*. By courtesy of the State Pedagogical Publishing House, Prague

49 Neandertal man. From J. Augusta and Z. Burian *Prehistoric Man*. By courtesy of the State Pedagogical Publishing House, Prague

64 Caricature of Darwin by 'Ape', from *Vanity Fair*, 30 September, 1871

65 Caricature of Huxley by 'Ape', from *Vanity Fair*, 28 January, 1871

96 Cave painting at Altamira. Photo by courtesy of the Commission Internationale d'Art Préhistorique, Bergerac

97 Cave painting at Rouffignac

112 Negative imprint of a hand surrounded by spots of pigment, Pech-Merle

113 Cave painting at Lascaux

128 The Lespugue Venus. Paris Musée de l'Homme

129 Woman with a bison horn, relief from Laussel. Musée d'Aquitaine, Bordeaux

144 Spearheads of the Palaeolithic period

145 Neandertal skull, frontal and lateral views. Broken Hill, Zambia. By courtesy of Dr M. H. Day and the Trustees of the British Museum (Natural History)

192 *Australopithecus africanus*, reconstruction painting by Maurice Wilson. By courtesy of the British Museum (Natural History)

193 Top: view of the Olduvai Gorge. Bottom: view of the excavation area where Zinjanthropus was found. Photos by courtesy of Dr M. H. Day

208 Zinjanthropus skull. Photo R.K. Campbell and Bruce Coleman Ltd

209 Louis and Mary Leakey comparing skulls at Olduvai. Photo Jen and Des Bartlett and Bruce Coleman Ltd

Index

Abevillian 112
Abel, Wolfgang 209
Acheulean 52, 112, 142, 143, 144, 164, 231
Adam 7
Alcalde del Rio 97, 105
Altamira, cave 96ff., 105, 122, 180
Andersson, J. G. 183, 184
Andrews, Roy Chapman 180, 183f., 185, 186, 187, 205, 258
anthropoid apes 16, 32f., 73, 74, 76, 88, 161, 168, 178, 186, 194, 197, 214, 243, 249, 250, 252, 258, 262, 269
ape-god 16, 18
Apidium, fossil primate 262
Arambourg, C. 198
Ardrey, Robert 9, 206, 215
Asselar (Sahara) 128
Atlanthropus mauritanicus 199
Aurignac, cave 90
Aurignacian (Brünn race) 112, 115, 117, 119ff., 123, 134
australopithecines (*see also* hominids) 206, 221, 222, 224ff., 231, 238, 239, 240, 241, 243, 244f., 256, 267, 276
Australopithecus 209, 214, 219, 221, 223, 224, 225, 226, 238, 241f., 246, 247, 262, 264, 266, 267, 268, 270, 277; A type 219, 222, 238, 242, 243, 276; P type 219, 228, 229, 234, 238, 241ff., 247, 276; *A. transvaalensis* 214
Australopithecus africanus (Taung child) 209, 214, 219, 222, 240, 242, 243; *A. prometheus* 242
Avebury, Lord, *see* Lubbock, Sir John
Azilian 113

baboon 222, 223, 225, 247
Balbal depression, Olduvai 232
Bandi, Hans-Georg 119, 122, 124f., 126
Bardon, Abbé 136
Barlow, G. W. 216f.
bear cult 137
Beer, Sir Gavin de 154
behavioural research 256
Berckhemer, F. 141
Beringer, J. B. A. 25f.
Berniche, M. 102, 103
Berthoumeyrou, Gaston 102, 103
Bibby, Geoffrey 92, 96, 128
bison 101, 109, 110
Black, Davidson 180, 185f., 188, 194, 201
Blumenbach, J. F. 37
Bohlin, Birgir 180, 185
Bölsche, W. 87
Bontius, Jacob 33
Boucher de Perthes, Jacques 51ff., 55, 88
Boule, P. M. 136, 180

Brain, C. K. 223, 236, 238
Brassempouy 121
Breuil, Henri, Abbé 99, 100, 103ff., 109, 111, 116, 119, 122, 123, 128, 180, 221
Broken Hill 155f.
Bronze Age 43f., 124, 126
Broom, Robert, 137, 183, 208, 211, 212ff., 219, 221, 223, 226, 228, 238, 247
Buckland, William 49, 87
Buffon, G. L. L., Comte de 34, 37ff., 41
burial places, *see* graves
Bushman 126, 266

Cabré, Aguilo, Juan 122, 124
Calapata (Spain) 122
Camper, Pieter 35, 37
cannibalism 129f., 136, 175, 189, 222, 270
Capitan, Louis 103, 104
Cárballo, Jesús 93
Carbon 14 dating 156f., 237
Carmel, Mount 138f., 144
Cartailhac, Emile 100, 103, 104f., 122
Castellón (Spain) 123
cattle breeding 106
cave bear 30, 50, 88
cave paintings 92ff., 97ff., 109, 113, 116f., 122
Chancelade race 117, 133, 134
Chellean 112, 241
chimpanzee 33, 71, 209, 214, 230, 249, 250f.; paintings 251, 252ff., 259, 266, 267
Chou-kou-tien 184, 185f., 188, 192, 193
Christy, Henry 88f., 99, 102
Clacton-on-Sea 144
Clark, Sir Wilfred le Gros 153, 223
Cogul 122f.
Cole, Sonia 232, 240f., 263f., 267
Combe Capelle, cave 133
Coppens, Yves 235
Cro-Magnon man 30, 90ff., 106, 112, 117, 128, 134, 137, 265
Curtis, G. H. 238
Curtis, Ludwig 15
Cuvier, Georges 38, 42f., 49, 50, 53, 56, 62

Daleau, François 101
Dames, Wilhelm 169f.
Dart, Raymond A. 206ff., 213, 221ff., 238
Darwin, Charles 18, 41, 50, 54, 57, 58, 59ff., 71, 73, 74, 76f., 88, 91, 92, 93, 132, 140, 150, 180, 246, 249, 274
Darwin, Erasmus 60

Dawson, Charles 149ff.
Deluge, *see* Flood
descent, theory of (*see also* genealogical tree) 73, 79f.
Diluvians, 'diluvial man' 29f., 34
Diluvium, *see* Flood
Dolní Vestonice 120, 121
Dryopithecus 88, 258, 259
Dubois, Eugène 166ff., 173, 176, 178, 201

Eckhardt, J. G. von 25
Ehringsdorf (Thuringia) 140
Elbert, J. 171
Engihoul, cave 48
Engis, cave 48
Eoanthropus dawsoni ('Dawn man') 145, 149
Eocene 269
Ericson, David B. 88
Esper, J. F. 30
Evans, Sir John 77
Evernden, J. 238
evolution, theory of 37, 40f., 51, 133, 187

Falconer, Hugh 66
Feldhof grottoes 55, 58
fertility cult 111
fire, use of 164
Fitzroy, Robert, Admiral 60, 62, 70
Flood, the 27, 43, 46, 49, 51, 87
fluorine testing 153, 156, 158
Fontéchevade 144
Font-de-Gaume, cave 104, 105f.
Frere, John Hookham 48
Fuhlrott, Carl 56, 57f., 134, 225

Gailenreuth, cave 30
Galapagos Islands 62
Garrod, Dorothy 138f.
Gaudry, Albert 258
genealogical tree 78ff., 138, 247, 260, 269; g. 'bush' 138, 191, 226, 247
Gervais, Paul 261
giant salamander 27f.
giants 193ff., 201, 228, 234
gibbon 161, 171, 249, 259
Gibraltar 135
Gieseler, W. 209
Gigantopithecus 191, 194, 197, 198; *G. blacki* 194
glaciations (*see also* Ice Age) 87, 88
Glowatzki, Georg 140, 144

'Goat's Hole' (Wales) 49
Goethe, J. W. von 36ff.
Goodall, *see* Lawick-Goodall, Jane van
Gorgon 16f.
gorilla 15, 17, 71, 75, 93, 155, 195, 196, 209f., 228, 235, 249, 259, 267
Gorjanović-Kramberger, K. D. 132
Granger, Walter 183
graves 15, 128f., 132f., 160
great apes, *see* anthropoid apes
Grimaldi caves 126ff.
'Grotte des Enfants', *see* Grimaldi caves
Günz-Mindel interglacial 88, 161

Haberer, A. 183, 184
Häberlein, C. E. J. 171, 173
Haeckel, Ernst 77, 79ff., 88, 140, 165, 166, 167, 170, 180, 249
Hamy, Ernest 135
hand-axes 43, 46, 48, 51, 94–5, 112, 143, 199
Hanuman (ape-god) 18
Harlé, Edouard 99
Hartmann, Daniel 160f.
Hauser, Otto 106f., 132f., 136
Heberer, G. 137, 141, 153, 227, 231, 243, 246f., 260, 262, 266, 268, 276
Heidelberg man 160ff., 200, 239, 264
Henslow, John Stevens 60
Hernandez-Pacheco 124
hominids (ape-men) 80, 82, 83, 134, 138, 150, 165, 167f., 171, 183, 198, 205, 209, 218, 220, 224, 228, 232, 238, 241, 247ff., 252, 256, 259, 260, 262f., 265, 267, 268
hominoids 262, 263, 265
Homo diluvii testis 29
Homo erectus 156, 190, 200f., 229, 231, 235, 239, 241, 247, 277; *H. erectus capensis* (*see also* Telanthropus) 241; *H. erectus leakeyi* 200, 241; *H. erectus pekinensis*, *see* Peking man; *H. erectus soloensis*, *see* Solo man
Homo habilis (*see also Australopithecus habilis*) 244, 245, 246, 247, 264
Homo heidelbergensis, *see* Heidelberg man
Homo sapiens 137f., 139, 141, 142, 145, 155, 161f., 201, 231, 240, 242, 277
Hooker, Sir Joseph Dalton 58, 66, 70
Hooton, Ernest A. 76
Hopwood, A. 258f.
Hoxne 48
Hürzeler, J. 261, 262
Huxley, Sir Julian 270f., 274
Huxley, Thomas Henry 58, 65, 67, 69ff.
hyenas 50, 225, 226

Ice Age 23, 30, 48, 53, 83, 87ff., 96, 98, 100, 111f., 115ff., 119, 121, 123, 124ff., 128f., 131, 132, 134ff., 138, 145, 159, 166, 191
interglacial periods 88, 140, 163
intermaxillary bone 35ff.
Iron Age 43–44
Irvine, William 18, 58
Isturitz (Pyrenees) 113

Jardin des Plantes (Jardin du Roi), Paris 34, 37, 40
Java man 132, 165ff., 170, 179, 191

Kedung-Brubus 166
Keith, Sir Arthur 149, 151, 209
Kelvin, Lord 54
Kent's Cavern (Torquay) 49f.
Kenyapithecus wickeri (*see also* Ramapithecus) 263, 264
Kesslerloch, cave 101
King, William 58
Kircher, Athanasius 23f.
Klaatsch, Hermann 129ff., 133, 136, 155, 160, 170
Knopf, Adolph 156
Koenigswald, G. H. R. von 82f., 147, 149, 152, 165, 170f., 173, 174, 175, 176f., 179, 183, 185, 186, 189, 191, 193ff., 201, 226, 227
Kohl-Larsen, Ludwig, 198, 234
Koro Toro 235
Kortlandt, A. 253, 256
Krapina 131f.
Kromdraai 216, 219
Kühn, Herbert 97, 106, 116f., 123f., 136

La Brea (Sicily) 126
La Chapelle-aux-Saints, cave 135, 136
La Madeleine, cave 89, 112
Lamarck, Jean-Baptiste 38, 39ff., 60
La Mettrie, Julien Offroy de 22
La Mouthe, cave 102, 103
La Naulette, cave 135
Lantian 193
La Rochette 133
Lartet, Edouard 88ff., 96, 258
Lartet, Louis 90f.
Lascaux, cave 114ff.
Laugerie Basse (Dordogne) 101
Lawick-Goodall, Jane van 252f.
Leakey, Jonathan 244
Leakey, Louis S. B. 9, 144, 159, 199, 226, 229ff., 236, 237, 238ff., 243, 245, 246, 252, 258, 259, 263, 266
Leakey, Mary 193, 232, 237, 259
Leakey, Richard 276, 277
Le Moustier 132, 133, 135
Le Placard (Charente) 113
Les Combarelles, cave 103
Les Eyzies 88, 102, 104
Lespugue, Venus of 119, 121
Les Trois Frères, cave 110, 111
Levantine art (of Spain) 122, 126
Ley, Willy 197
Libby, Willard F. 158f.
Linné, Carl (Linnaeus) 31f., 34, 37, 71
Lorenz, Konrad 254
Lubbock, Sir John 46, 100
Lyell, Sir Charles 53ff., 58, 59, 63, 66, 71, 88

Mackintosh, R. 223, 224
Magdalenian 112, 113, 115, 121, 123, 124, 144

Major, Forsyth 261
Makapansgat (Makapan valley) 206, 222, 227, 241, 246, 247
Malthus, Thomas R. 63f.
mammoth 23, 48, 50, 51, 88, 91, 109
man-apes, *see* anthropoid apes
marine lizards 61
Maringer, J. 119
Marsoulas, cave 101
Marston, A. T. 142
Maška, J. K. 135
Mauer, cemetery at 160f.
Max, Gabriel 82
Mayer, Robert 56
McEnery, J. 49, 54
Meganthropus (*see also* Australopithecus) 196, 197, 201, 229, 242; *M. africanus* 234; *M. palaeojavanicus* 194, 195, 228
'megatherium' 42
Merian, M. 23
Merriam, J. C. 177
Mindel-Riss interglacial 88, 140, 142
Miocene 251, 258, 264
'missing link' 32, 33, 77, 88, 162, 166, 169, 180, 183, 188, 205, 208, 211, 214ff., 219, 228
Modjokerto (Java) 176f., 195
Monte Bamboli 260ff.
Monte Circeo 130
Montelius, O. 100
Monte Pellegrino (Sicily) 125
Moore, Ruth 151
Morant, G. M. 144
Morris, Desmond 9, 250, 251f., 256
Mortillet, Gabriel de 92, 106, 112
mountain ape, *see* Oreopithecus
Mousterian 112, 133, 134, 135
Muller, H. J. 272f., 274, 275

Napier, John 232, 246, 268
natural selection, *see* selection
Neandertal(er) 30, 56f., 77, 131, 132, 133, 134ff., 144, 155, 160, 168, 175, 201, 265
Neander valley 55, 56, 58
Nehring, Alfred 168
Neolithic Age (New Stone Age) 46, 140
Ngandong 173, 174, 175
Ngorongoro 232
nomadic hunters 90f., 106, 129, 136, 228
'Nutcracker' man 201, 228, 232

Oakley, Kenneth P. 153, 154, 156, 238, 241
Obermaier, Hugo 106, 119, 123, 128, 180
Oken, Lorenz 75f.
Oldoway man 238, 241, 245f.
Old Stone Age, *see* Palaeolithic Age
Olduvai Gorge 193, 199, 230ff., 236, 238, 241, 244, 245, 247, 252, 264, 276
Oligocene 260, 262, 269

Oppenoorth, W. F. F. 171, 173, 174
orang-utan 33, 35, 228, 249, 250, 267
Osborn, H. F. 213
Oreopithecus 261, 263; *O. bambolii* 261

Pair-non-Pair, cave 101
Palaeanthropus 199
Palaeolithic Age 112, 113, 124, 128f.,
    137, 232
Paranthropus crassidens 228, 242; *P.
    robustus (see also Australopithecus ro-
    bustus)* 219, 242
Pavilland, 'Red Lady of' 49
pebble tools 164, 223, 231, 232, 240,
    241, 242, 256
Pech-Merle, cave 107f.
Pei Wen-chung 180, 186, 192, 201
Peking man 138, 179, 182, 186ff., 197,
    198, 232, 241, 264
Pengelly, William 49f., 51, 55
Peninj (East Africa) 235
Peyrony, Denis 103f.
'picture stones' (Würzburg fake carv-
    ings) 25f.
Piltdown man, 145, 149ff., 156, 160,
    209, 238
Pithecanthropus *(see also Homo erectus)*
    146, 156, 167ff., 172, 175, 177f., 189,
    195; *P. alalus* 80, 83, 165; *P. erectus*
    165, 167f., 197; *P. robustus* 195, 197
Platter, Felix 23
Pleistocene 88, 119, 122, 137, 159, 173,
    198, 236, 247, 252
Plesianthropus transvaalensis *(see also
    Australopithecus africanus)* 219, 220,
    221, 242, 266
Pliocene 80, 187, 198, 252, 258, 264
potassium-argon method of dating 159,
    237, 241, 264
Praeneste 15
pre-Neandertalers 52, 139f., 142, 144,
    156, 175
pre-sapiens man *(see also hominids)* 52,
    141, 144, 156, 160, 164, 191
pre-Zinjanthropus 244
primates 30, 71, 182, 249, 252, 266, 269
Proconsul *(see also Dryopithecus)* 138,
    259f., 262, 263
Proliopithecus haeckeli 260

Quaternary era 105
Quinzano 144

radiocarbon testing *(see also Carbon 14
    dating)* 156, 158f.
Ramapithecus 264, 265f., 268
Ramsay, Sir Andrew 52
Ranke, J. 162f., 170
Reck, Hans 230, 231
reindeer 50, 89, 101
Rensch, B. 251

Reynolds, Frankie and Vernon 253
Rhodesia man 145, 155, 156, 231, 265
Rigollot, Dr, of Amiens 52
Riss glaciation 88, 144
rites, initiation, hunting and fertility
    111
Rivière, Emil 102f.
Robinson, John T. 215, 216, 221, 223,
    228, 239
rock-shelters 89
Roderique, I. 25, 27
Rösch, J. 161
Rouffignac, cave 109
Rousseau, Jean-Jacques 37
Rudolf, Lake 276
Rusinga Island 258f., 260

Saint-Gaudens 88, 258
Saldanha man 156
Sangiran 177f., 179, 195
Sautuola, Don Marcelino de 96ff.
Sautuola, María de 97, 100, 105
Schaaffhausen, H. 56, 57
Scheuchzer, J. J. 27ff., 34, 49
Schlosser, Max 183, 184
Schmerling, P. 48, 54
Schoetensack, Otto 161, 163
Schwalbe, Gustav 134, 170
Sedgwick, Adam 66
selection 63, 65, 66, 73
Selenka, Emil 171
Selenka, Margarete Lenore 171
Serengeti plain 230f.
Shaw, George Bernard 274
Shipka cave 135
Sigrist, K. 140
Simons, E. L. 198, 265
Simpson, G. G. 242
Sinanthropus 185, 187, 189
Sloane, Sir Hans 28
Smith, Elliott 149, 173
Smith-Woodward, Sir Arthur 149,
    151f., 209
Smuts, Jan Christiaan 213, 220
Solo man 174, 175, 197
Solo River 165, 167, 171, 173, 178, 201
Solutrean 112
Steinheim 140f., 142, 144, 241, 265
Sterkfontein 205, 214f., 216, 217, 218,
    219, 221, 223, 228, 238, 264
Sutcliffe, A. J. 226f.
Swanscombe 142ff., 265
Swartkrans 228, 229, 239, 241
Szombathy 119

Taung *(see also Australopithecus africanus)*
    205, 207, 209f., 213f., 216, 219, 220
Teilhard de Chardin, Pierre 152f., 180ff.,
    185, 187, 189f., 221
Telanthropus capensis 239, 241
Tembrock, G. 254
Terblanche, Gert 216ff.
Ter Haar, C. 173, 175

Ternifine 199
Terra, Helmut de 180f., 190f.
Tertiary (era) 27, 79, 80, 88, 162, 173,
    180, 198, 221, 226, 230, 236ff., 252,
    260, 261
Thomsen, C. J. 43f., 46
Thomson, Sir William; *see* Kelvin, Lord
'Three-Age system' 43f., 46
Tobias, P. 246
toolmaker, man as 187, 241f., 268
Trinil 165, 166, 171, 178, 179
Tuc d'Audoubert, cave 110
Tulp, C. P. (Nicolaus) 33, 34
Tyson, Edward 33

Universal Deluge, theory of 27
uranium clock 54, 156ff.

Valltorta (Spain) 123
Ventimiglia, *see* Grimaldi caves
Vértes, László 163, 165
Vértesszöllös 163
Vesalius, A. 21f.
Vézère, valley of the 98f., 112, 132
Vilanova, Juan 96, 99
Villafranchian period 221
Virchow, Hans 140
Virchow, Rudolf 57, 80f., 100, 134, 140,
    165, 168, 169
Vogt, Karl 74, 76, 79
Voltaire (Arouet, François Marie) 38

Wadjak (Java) 166, 197
Wallace, Alfred Russel 65, 66, 73
Weidenreich, F. 175, 178, 185, 193,
    196f., 201
Weiner, J. S. 153
Weinert, H. 141, 198, 209, 222f.
Wilberforce, Samuel 70
Willendorf, Venus of 119
Windmill Hill 50
Wolfe, John B. 254
Wollin, G. 88
Woodward, John 27
woolly rhinoceros 50, 88, 109, 117, 135
Worsaae, J. J. A. 46, 51
Würm glaciation 88, 112, 129, 139
Wu Yu-kang 192, 197–8

Young, C. C. 180, 185
Young, R. B. 208

Zdansky, O. 183
Zeuner, F. 158
Zinjanthropus boisei 193, 230, 234ff.,
    242
Zuckermann, Sir Solly 249f.